# TWIGGY

# TWIGGY

THE HIGH-STAKES LIFE OF ANDREW FORREST

## ANDREW BURRELL

Published by Black Inc.,
an imprint of Schwartz Media Pty Ltd
37–39 Langridge Street
Collingwood VIC 3066 Australia
email: enquiries@blackincbooks.com
http://www.blackincbooks.com

Copyright © Andrew Burrell 2013
Andrew Burrell asserts his right to be known as the author of this work.

ALL RIGHTS RESERVED.
No part of this publication may be reproduced, stored in a retrieval system, or transmitted in any form by any means electronic, mechanical, photocopying, recording or otherwise without the prior consent of the publishers.

The National Library of Australia Cataloguing-in-Publication entry:

Burrell, Andrew John.

Twiggy : the high-stakes life of Andrew Forrest / Andrew Burrell.

ISBN: 9781863956208 (pbk)

Includes index.

Forrest, Andrew. Businessmen – Western Australia.
Executives – Western Australia. Mineral industries – Western Australia.
Mines and mineral resources – Western Australia.

338.209941

## CONTENTS

| | |
|---|---|
| *Author's note* | ix |
| *Prologue* | 1 |
| 1. The Forrest Legacy | 5 |
| 2. Minderoo | 21 |
| 3. Fisticuffs and Failures | 34 |
| 4. The Wild West | 47 |
| 5. Sin City | 61 |
| 6. Dreams in the Desert | 76 |
| 7. The Comeback Kid | 103 |
| 8. Burke's Backyard | 121 |
| 9. To Get Rich Is Glorious | 137 |
| 10. The Great Escape | 155 |
| 11. China Crisis | 169 |
| 12. Taxing Times | 191 |
| 13. The Intervention | 215 |
| 14. Doing God's Work | 242 |
| 15. Wealth and Power | 264 |
| *Acknowledgments* | 275 |
| *Notes* | 277 |
| *Index* | 299 |

*for Rebecca, Sebastian and Lily*

## AUTHOR'S NOTE

This is an unauthorised biography. I asked Andrew Forrest to co-operate with my research, but he declined. He also instructed his family members and some of his close friends not to talk to me. Forrest's opposition to this book greatly surprised many of those who know him well. It also surprised me. Since 2006, I have covered the resources industry and the rise of Fortescue Metals Group as a journalist in Perth for both the *Australian* and the *Australian Financial Review*. Senior executives at Fortescue have told me they believe my reporting has been fair and accurate.

Forrest's cooperation would have led to a different book, but not necessarily a better or more reliable one. His decision not to talk forced me to interview dozens of people whom otherwise I might never have called. Many of them provided sharper insights into his past than Forrest might have been prepared to do himself. And, luckily, some of his close friends and associates ended up defying his wishes and speaking to me at length.

Despite Forrest's non-cooperation with this book, I felt a heavy responsibility to be fair; wherever possible, I have sought to present his side of the story.

# PROLOGUE

*September 11, 2001: hotel room, Manila.* Andrew Forrest sits alongside a convicted drug dealer, glued to the television as the World Trade Center crashes to the ground in Manhattan. His own world is collapsing back in Australia and he has flown to Manila because he needs cash. Forrest has been booted out of the mining company he launched, his investors are baying for blood and he needs millions just to hold onto the family home. Within months, he will have left Australia to live in self-imposed exile, his ego bruised and his dreams cruelled. Most in the establishment see him as a cowboy and hope they have seen the back of him.

*March 12, 2012: Westminster Abbey, London.* Andrew Forrest stands before Queen Elizabeth II, Prince Charles, heads of government and hundreds of other VIPs to explain his quest to end indigenous disadvantage in Australia and rid the world of slavery. He is the only Australian invited to speak at the Commonwealth Day ceremony held in the imposing Gothic church. By now, he is respected as a generous philanthropist and backed by a multibillion-dollar fortune, thanks to the spectacular success of his Fortescue Metals Group.

How did he do it? How did Andrew "Twiggy" Forrest transform himself from a corporate pariah into a humanitarian courted by the global elite? How did the desperate debtor of 2001 become Australia's richest man by 2008? How could a country boy with an embarrassing stutter become the slickest salesman in Australian business? And how did a church-going Christian become a swashbuckling tycoon who frequently pushes against the barriers of acceptable behaviour?

The great Australian mining boom of the early twenty-first century is replete with tales of towering ambition and colossal wealth, but the story of Andrew Forrest's unlikely journey is perhaps its most enthralling. It is a study of bravado, guile, folly, adversity, triumph and vindication.

At school, Forrest was a rebellious, stick-thin kid who was quick with his fists. Later, his business ethos was shaped in the get-rich-quick culture of Perth's stockbroking scene of the 1980s, when crooks like Alan Bond and Laurie Connell were on the make. By 2001, his company, Anaconda Nickel, was in dire straits and his reputation was shot to pieces. Seven years later, he had mined the fastest sharemarket fortune in Australian history and was worth $10 billion. Now he is remaking his image yet again, befriending world leaders as he promises to give away most of his money to charity.

Among the tribe of Australian miners who have become billionaires in the resources boom, Forrest is the most substantial business figure. Nathan Tinkler rose and fell in the blink of an eye. Clive Palmer claims to be worth billions but has never actually mined a deposit and seems more dedicated to self-promotion. Gina Rinehart may be worth more on paper than Forrest, but in twenty years of running her own company she has yet to build a mine on her own and may struggle to do so, given her aversion to risk and her well-documented idiosyncrasies.

Forrest is a deeply polarising figure. At one end of the spectrum, people worship him for the brilliant and unconventional way he built one of the biggest private projects ever undertaken in Australia. They point to his mission to improve the lives of indigenous people and fight modern slavery as evidence of his underlying decency.

At the other extreme, people view him as a hyperbolic fool who can't be trusted to deliver on his promises and is motivated by greed. Four judges in four separate court cases have questioned his truthfulness. "An idiot who got lucky," is how one prominent Perth business leader describes him.

The story of the mining boom is writ large in the rise of Andrew Forrest. And nowhere is the impact of the China-fuelled economic miracle more palpable than in his home town of Perth. When Forrest sketched out his plan to launch Fortescue Metals Group while sitting at his kitchen table in 2003, the median house price in Perth was $200,000 and iron ore was selling for just $US30 a tonne. Five years later, the median house price had more than doubled to $450,000 and iron ore was worth $US200 a tonne as the urbanisation and industrialisation of China became a seismic economic event.

Back in 2003, resource companies were shipping commodities worth $26 billion out of Western Australia – a healthy industry by any measure. But ten years later, the sector's value had reached $106 billion. The glamour commodity is now iron ore, the steel-making ingredient that lies beneath the rust-red hills of the Pilbara. It is worth $60 billion annually to Australia – that's a staggering $165 million worth of dirt that is being dug up each day in the Pilbara. Iron ore now plays a big part in determining whether the federal and state budgets are in surplus or deficit.

Forrest is at the forefront of this economic revolution because Fortescue was the first company to challenge the supremacy of iron ore miners Rio Tinto and BHP Billiton in the Pilbara. In 2003, Fortescue was little more than a thought bubble in Forrest's hyperactive brain. Within five years, it was the world's fourth-biggest iron ore exporter.

Forrest has succeeded because he is an entrepreneur in the classic sense. "To understand the entrepreneur," renowned psychoanalyst

Abraham Zaleznik told the *New York Times* in 1986, "you first have to understand the psychology of the juvenile delinquent." Zaleznik meant that the hallmark of a great entrepreneur is the drive for autonomy that bespeaks rebelliousness and fearlessness in the face of risk.

As this book will explain, Forrest has always had a rebellious streak and doesn't believe that conventions apply to him. The rules are there to be pushed and prodded until he gets his way. He is afraid of nothing and somehow wills himself to believe that he will never fail. But he is also highly charismatic and socially charming, inspiring a cult-like loyalty and optimism in his followers.

The great Austrian-American economist Joseph Schumpeter could as easily have been describing Forrest's character when he wrote, more than a century ago, that the best entrepreneurs are propelled by "the will to conquer, the impulse to fight, to prove oneself superior to others". The ruthless manner in which Forrest fought his way to the top is a modern Australian case study in entrepreneurial success.

# 1.

# THE FORREST LEGACY

*He's seriously got whatever it was that drove Alexander and John Forrest and David Forrest to go into the deserts of the middle of the country. It's in the history, it's in the blood.*
—JANIE HICKS, Andrew Forrest's sister

When Judy Forrest discovered she was pregnant in 1961, her first reaction was to plan for a miscarriage. Life on a remote sheep station in the Western Australian outback was demanding enough with two small kids and a husband gone from dawn to dusk. For Judy, the thought of having a third child was horrifying. So she decided to ride her horse bareback across the rugged Pilbara terrain every day for a week, with the aim of aborting the pregnancy. When that didn't work, she jumped off the roof overlooking the homestead's tennis court three times a day for seven days, but without success. Friends advised her to hit the gin bottle as well, but she drew the line at that.

Several months later, on 18 November, Judy gave birth to a healthy baby boy, John Andrew Henry Forrest. She confided in her youngest son from an early age that she had wanted to end the pregnancy and told him he must be a tenacious type to have survived in the womb. It may have been a calculated ploy by the tough-as-nails Judy to instil in her youngest son a sense of self-belief and drive that she believed would take him far in life. According to the small group of friends who have heard this story from Forrest over the years, it also implanted in the boy's brain the overwhelming need to impress his mother. "So much of what he does is proving to his mother that he's worth it," says Warwick Grigor, Forrest's former business partner. "It really is a big motivating factor."

Young John carried another weighty burden from birth: his name. Being named after Sir John Forrest, his great-great-uncle and Western Australia's most famous historical figure, was simply too much to bear for the boy, and so he decided he'd be known by his second name, Andrew. Despite this, many now believe that Andrew Forrest, the entrepreneur, has been inspired to take extraordinary risks and pursue grand dreams by seeking to follow in the footsteps of Sir John, who was the state's first premier and most celebrated pioneer. Warwick Grigor, who knows Forrest better than most, suggests the businessman's success is driven by a combination of living up to the Forrest family legacy while subconsciously trying to prove to his mother that he deserved to be born in the first place.

Over the years, Forrest has rarely missed a chance to cite the deeds of Sir John in the self-promotional spiels he delivers when wooing a financier, pitching his vision to a politician or signing up a customer. Often he has worked at a desk underneath a framed portrait of the burly, bearded image of "Uncle John". Even his home address contains potential clues to his family inspiration. For the past thirteen years, Forrest has lived in Perth's salubrious beachside suburb of Cottesloe at the pinnacle of John Street, which runs parallel to Forrest Street – both leafy thoroughfares were named after Sir John Forrest. Perhaps Forrest's decision to buy a house in this location was a coincidence. If so, it's an extraordinary one. For his part, Forrest has mostly rejected the notion of himself as a latter-day incarnation of Sir John. Yet the words of his family members and others who've observed him up close have tended to blunt his denials.

"I think, unconsciously perhaps, Andrew would like to emulate his great-great-uncle," said Forrest's father, Don, when asked whether his son was motivated by the deeds of Sir John. Forrest dismissed such talk in a rare interview on the topic with Perth journalist Mark Drummond in 2007. "We can never escape who we are," he said. "We are the opportunities, the education and the genetics we were given. But as far as it [the Forrest name] being a motivator to me, that's entirely misguided."

By 2011, however, after his goal of building an iron ore mine in the vast red dirt of his Pilbara homeland had been realised, Forrest seemed more willing to accept that he had been inspired by Sir John and his sidekick CY O'Connor, the brilliant engineer who designed a revolutionary water pipeline to the Goldfields but committed suicide after being accused of corruption. "I think those two guys probably did have an impact," Forrest said. "To think, okay, if you really believe in something yet you're heavily ridiculed for it, well, just cop it. Don't make the mistake which CY did, which is to let it all get to you ... I was fascinated by [O'Connor], fascinated by the fact he was told it's absolutely impossible to create a dam at what is now the most beautiful Mundaring Weir [outside Perth]. He was told it was completely impossible to pump water up a hill for 600 miles, and I love the fact that John Forrest believed so much in him and backed him completely."

Graeme Kirke, Forrest's former colleague at Perth stockbroking firm Kirke Securities, believes Forrest's raw determination, obsessiveness and zeal are actually embedded in his DNA. "Do I think it [the family legacy] is on his shoulder? No, I don't – I think it's innate," he says. Forrest's sister, Janie Hicks, is also a believer in the nature, rather than nurture, theory when analysing her famous brother's personality. "He's seriously got whatever it was that drove Alexander and John Forrest and David Forrest to go into the deserts of the middle of the country," she said in 2005.

"It's in the history, it's in the blood." Forrest's older brother, David, agreed: "Andrew will probably go down as a very significant part of the 21st-century version of John."

To fully understand Andrew Forrest, we need to delve into the history of Western Australia and examine the exploits of Sir John. There are, in fact, plenty of similarities between Andrew "Twiggy" Forrest, the entrepreneur of the late twentieth and early twenty-first centuries, and John Forrest, the explorer, politician and statesman of the late nineteenth and early twentieth centuries. Twiggy has said his business motto today – of taking bold but calculated risks – is essentially the same as the philosophy Sir John employed in the desert more than a century ago. John Forrest survived his three dangerous expeditions into the continent's interior by never venturing so far from a waterhole that it would be impossible to return. It's a fitting analogy because Andrew Forrest has strayed from the metaphorical waterhole a few times in his colourful career. And so far, like Sir John, he's always made it back – sometimes bloodied yet still alive.

Like John Forrest, Andrew has ventured into frontier territory and built some of the essential wealth-generating infrastructure of his era. Sir John borrowed six times the annual WA government budget – an extraordinarily risky amount, equal to a premier taking out loans of more than $100 billion today – to build the Goldfields water pipeline and other nation-building works at a time when the colony was beginning to grow rapidly, thanks to the discovery of gold at Kalgoorlie and Coolgardie. Andrew Forrest has shown a similar fondness for debt, raising billions of dollars to build the Murrin Murrin nickel project north of Kalgoorlie in the 1990s and, more recently, the Fortescue iron ore mines of the Pilbara. John Forrest, a "bulldozer of a man" in the words of historian Geoffrey Bolton, was the pivotal figure of the state's first mining boom in the 1890s. Andrew Forrest, by dint of his wealth and activism, is the poster boy of the latest one. Both rose to the top with boundless energy, habitual optimism, self-belief and an ability to inspire others. Both were gamblers. Luck also played a part, for both Forrest men were in the right place at the right time.

Born in 1847 near the WA town of Bunbury, south of Perth, John Forrest was the son of indentured Scottish migrants who had arrived in the colony in 1842 with virtually nothing, spending their first night huddled in a tent. John was a tall, strong youth who sprung to fame at the age of twenty-one by leading a daring expedition into the heart of Australia in search of the missing explorer Ludwig Leichhardt. Leichhardt's party had last been seen twenty years earlier, about 2400 kilometres east of the most easterly sheep station in Western Australia, while attempting to cross the continent from east to west through the interior.

This was an era in which the colony of Western Australia, settled only in 1829, was still almost totally isolated from the rest of the continent and was economically backwards in most respects. Forrest, a surveyor by profession, left Perth in April 1869 with five other men, sixteen horses and several dogs. After traversing more than 3000 kilometres of desolate country in the most remote interior of Western Australia, he failed to find any trace of the Leichhardt party. Nor did he find, as he had hoped, an inland Garden of Eden, a river system or even any decent pastoral land. As John Forrest's biographer Frank Crowley noted, the most significant outcome of the expedition into the "Great Lone Land" was the confidence Forrest developed in his own abilities and the respect he inspired in both his subordinates and superiors.

Yet there was another benefit of that first mission: after Forrest's compass became affected by what he assumed was the presence of underground minerals, he recommended that geologists be sent to what would become known throughout the world as the WA

Goldfields. It was in this desert region, a century later, that Andrew Forrest would forge another new frontier by establishing a $1-billion nickel operation.

John Forrest set off on another dangerous expedition in 1870, this time aimed at surveying the area between Western Australia and South Australia – a region along the coast of the Great Australian Bight then known as No Man's Land. Earlier, the explorer Edward John Eyre had traversed the same coastline but had almost died of thirst and, in a desperate dash for survival, had taken few detailed notes. Forrest chose his twenty-year-old brother, Alexander, who had also qualified as a surveyor, as second-in-command, and Aboriginal tracker Tommy Windich as his guide. In total, there were six men, sixteen horses and several dogs. The group survived on damper and salt pork, slept in the open for more than three months and suffered from unbearably painful feet after walking vast distances in the hot, fly-plagued country. Finding enough water for themselves and the horses was a daily struggle. Once they were forced to make a 250-kilometre dash to a waterhole with just one litre of water each per day. A few weeks after his twenty-third birthday, Forrest arrived in Adelaide to a hero's welcome. While the tangible results of the expedition were again limited, Forrest had successfully led the first west-to-east crossing of Western Australia. The mission also helped bridge the gap that was finally closed seven years later with the completion of the telegraph line connecting Perth to London, via Adelaide.

But the mysteries of the vast interior still nagged at Forrest and the British colonists of the era. Again backed by Alexander Forrest and Tommy Windich, Forrest began his longest and most daring journey in March 1874 – a 4000-kilometre transcontinental expedition that started in Geraldton, 400 kilometres north of Perth, and went straight through the Australian centre. According to writer Cyril Ayris, the mission earned the group the sort of fame "that in our day comes to young men and women as successful Olympic athletes or tennis stars". Ayris wrote that this was Forrest's most difficult trip of all.

> Water became scarce and grass gave way to spinifex which made walking difficult and cut the horses' flanks. The terrain was an endless sea of corrugated sand hills. The sun beat down on them and the glare of the sand dazzled them. This was terrible country – worse than any of them had seen.

The party were twice attacked by naked, spear-throwing Aborigines, possibly because they had camped inadvertently on a sacred site. As the group trudged up and down sandhills, Forrest began to pray to his "Creator" to guide him to water. One of the men suffered scurvy and his feet became so badly swollen that he could hardly walk. By September, as they followed the dry bed of a river in the northern part of South Australia, the men were subsisting on nothing but flour porridge three times a day. John Forrest had lost ten kilograms in weight. Finally the group reached the overland telegraph line that ran through the centre of the continent from Darwin to Adelaide. Upon reaching Adelaide in early November, they were given another civic reception, with big crowds lining the streets. Again, Forrest publicly regretted that the land he had traversed was largely useless and never likely to be settled. But Forrest's fame had grown immeasurably. He travelled to London, where he was presented to Queen Victoria, and upon his return was promoted to deputy surveyor-general of Western Australia at the age of just twenty-eight.

Forrest never took unavoidable risks and no man died on any of his three expeditions. This was a time when many fellow explorers had perished in the Australian interior because they pushed ahead too quickly without ensuring a supply of water. By all accounts, Forrest had great personal attributes. He was physically strong and able to survive for as long as forty-eight hours without food and with only the smallest amount of water. Yet Sir John's name is not readily associated with the nation's greatest explorers. Historian Geoffrey Bolton believes that outside Western Australia his achievements are largely ignored in favour of "tragic incompetents such as Burke and Wills".

In 1876 John Forrest married a refined young woman, Margaret Hamersley, and the couple moved into a magnificent house called The Bungalow in the centre of Perth, which had been bequeathed to Margaret by her late father. Forrest, the son of a humble miller, had married into the city's elite. John and Margaret were treated as WA royalty and The Bungalow was at the centre of much of Perth's social and political activity. To their deep regret, however, the couple could not have children, possibly because Margaret had suffered injuries in a horseriding accident as a teenager. Once married, Forrest returned to his day job as a surveyor, spending months at a time away on work that helped formed the basis for the eventual mapping of the entire state of Western Australia.

If Forrest's achievements had ended at this point, he would still be remembered as one of Western Australia's most important citizens. But he was about to embark on a spectacularly successful career in state and federal politics, which would take him to within one vote of the prime ministership. At age thirty-five, Forrest had become surveyor-general and a member of Western Australia's executive council – the first locally born man to be admitted to the colony's highest governing body. It did not take long for Forrest, always a man of strong opinions, to clash with the colonial governor of the day, Frederick Broome, whom Forrest accused of ruling in a despotic style. But Forrest established a reputation as a shrewd political tactician and was regarded by the public as a "local boy made good".

When Britain granted Western Australia the right to self-rule in 1890, John Forrest was elected unopposed to the seat of Bunbury in the Legislative Assembly. He then threw his hat in the ring to become premier, gathering supporters among fellow members of parliament and lobbying the governor at that time, Sir William Robinson, to give him the job. Crowley observed that the hugely energetic Forrest was desperate to become Western Australia's first premier.

> Why did he seek the premiership so ardently? The answer can only be found in the character of the man as it had been influenced by his career. He was without doubt a born leader, a man with a remarkable degree of self-confidence, who, unlike many gifted in that manner but not so fortunate, had been able to take advantage of situations best suited to his temperaments and his natural talents. He had an appetite for leadership and power, and a temperament well-suited to decision-making.

After being named the state's first premier, Forrest quickly employed CY O'Connor, who had twenty-five years' engineering experience in New Zealand, as his chief engineer to build a harbour at Fremantle and find a way of transporting water to the Goldfields. The appointment was one of his best decisions. Public works were urgently needed because the British government had been unwilling to approve spending in the distant colony. Across the vast state, railways, ports, jetties, lighthouses and town halls would be built by the Forrest government in a frenzy of activity. When O'Connor first cabled Forrest to ask what his new job would entail, Forrest simply replied: "Railways, harbours, everything." The water pipeline, in particular, was one of the world's great engineering feats. Water was dammed outside Perth at Mundaring Weir and piped uphill for more than 550 kilometres to Kalgoorlie.

To this day, it is a lifeline for mining towns perched on the edge of the desert.

In both its far-sightedness and scale, the Forrest government's provision of infrastructure is arguably unparalleled in Australian history. The town of Southern Cross, 370 kilometres east of Perth, had a railway by 1894, just four years after it was gazetted. Kalgoorlie and Coolgardie were connected by telegraph by 1894 and had their railway two years later. These were towns experiencing a population explosion as a result of the gold discoveries that were making news worldwide. Perth itself experienced huge growth and was catapulted from the ranks of country town to one of the top ten cities of Australia by the turn of the century. Western Australia's population soared from 100,000 to 240,000 in the decade to 1904, a period in which unemployment was high and industrial action common in the eastern states. Unlike today's mining boom, which has attracted more international than interstate migrants, most new settlers to Western Australia in the 1890s came from the economically depressed states of Victoria, New South Wales and South Australia.

For better or worse, it is difficult to imagine governments of today acting so rapidly with such huge amounts of taxpayers' money. But Forrest had the political courage to stare down his many critics and see projects through to completion, even when accused – at home and abroad – of corruption or incompetence. London's *Financial Times*, in a report on the economic miracle that was occurring in the colony under Forrest, wrote of the "waste and extravagance which one sees throughout Western Australia". But Forrest, the politician, took political credit for the public works he was building, telling parliament that future generations would thank his government. By this time his authority had been enhanced by a knighthood – he was the first native-born Western Australian to receive the honour. O'Connor, a more reclusive and sensitive man, shot himself while riding his horse into the waves at a beach near Fremantle in March 1902, aged fifty-nine, a year before the pipeline was commissioned. Weeks before his suicide, an article in Perth's *Sunday Times* newspaper had accused him of "gross blundering" and "reckless extravagant juggling with public funds".

Another of Forrest's great legacies is his guidance of Western Australia into the federation in 1901 despite strong opposition from within his rural electorate, his ministry and even his friends. Throughout the 1890s Forrest was a member of the national meetings that framed the Australian Constitution, and he argued repeatedly for a strong upper house to give the less populous states a voice. He also lobbied the eastern premiers – largely unsuccessfully – for a number of concessions for Western Australia to enter the federation. Forrest called a referendum in 1900 and the margin was two-to-one in favour of federation, with women voting for the first time. The mining boom in the Goldfields and the arrival of tens of thousands of "t'othersiders" from the east had boosted public support for Western Australia to join the commonwealth. But many of his former supporters never forgave Forrest for championing the "yes" vote, and he would be blamed for the supposed negative effects of federation on the state's economy and finances for the next thirty years. Many opponents of federation soon became firm advocates of secession – a fringe movement that survives in the west to this day.

As premier, John Forrest is credited with many social reforms, including extending voting rights to women. But in truth Forrest was nowhere near as visionary in his social outlook as in his economic policies. His attitudes to women were staunchly conservative, which was hardly unusual for the era. He was pleased when an early motion to allow female voting was defeated in parliament, believing that it would lead to women sitting in parliament and neglecting their home duties. Forrest told parliament in 1897 that men were created to be breadwinners and women were intended to be "the comfort and solace of our homes". By 1899, however, he'd had a sudden change of heart. It is thought that his wife, Margaret, who by that time had joined the Suffrage League, inspired the premier's sudden enthusiasm for the cause.

Forrest's attitudes to Aborigines were similarly unenlightened. Historians Bob Reece and Tom Stannage concluded that Forrest and his colleagues were "locked into and promoted an ideology of development which had racism at its heart". While it seems clear that Forrest had a sympathy for individual Aborigines who had helped him as a surveyor and explorer, he had no vision of a future for indigenous people in a white society and viewed the surviving Aborigines in settled areas as little more than a public nuisance. He believed it was the government's duty to help indigenous people in need "until the race died out". But as an investor in sheep stations, Forrest needed Aborigines as a source of cheap labour, and in 1893 he called for restraint by those whites who would regularly shoot Aborigines caught stealing food or cattle. "It's all very well for us to be incensed against these native outrages, but we must remember this: they are not all bad," he said. "How would we like to be shot at when we had done nothing wrong? Those who do the mischief deserve punishing, no doubt, but this sort of random retribution would kill both the innocent and the guilty … We must endeavour to civilise them by degrees. I must not, in the position that I am in, do anything or sanction anything that will lead to the impression that an indiscriminate slaughter of blackfellows will be tolerated or allowed by the government of the colony."

Unsurprisingly, his attitudes to foreigners were hardly progressive either. Forrest told parliament he did not want any Asians to settle permanently in the colony. "There are millions of them, and if we do not place some restrictions on them they will overrun the country, and, instead of being a British country, this will be an Asiatic country," he said. Forrest introduced an Immigration Restriction Act in 1897, which established a dictation test aimed at excluding the Chinese from living in Western Australia, and he was proud that his government had not allowed Asians to be issued with mining rights. This sort of unconcealed racism was common, of course, at the time. Frank Crowley believes Forrest "took with him to his grave this compound of social snobbery, laissez faire capitalism, sentimental royalism, patriotic Anglicanism, benevolent imperialism and British racial superiority".

By 1900 John Forrest had been joined in parliament by two of his brothers, Alexander and David. The three of them owned the vast Minderoo sheep station, which David Forrest had managed since the 1870s. John and Alexander Forrest, in particular, became the target of gossip that they took advantage of their government work to acquire big landholdings. But there was no evidence of corruption, certainly not on John's part. "He [John] used his own money to finance expeditions, didn't charge travelling expenses and, as an early member of parliament, was not paid," wrote Forrest family historians Alison and Dinee Muir. "He refused to take a parliamentary pension as he said the country could not afford it and he had enough money of his own."

Alexander, however, proved to be exceptionally gifted at seizing the lucrative business opportunities that were emerging in the fast-growing economy, accruing interests in pastoral stations, gold mines, newspapers, the timber industry, butchering and cattle shipping – even as he sat in parliament. He was the first genuine entrepreneur in the Forrest family and would remain its most successful until his great-grand-nephew became Australia's richest man in 2008. According to the Forrest family's historians, Alexander Forrest was a "prolific investor and played the stockmarket with zest". After leading a gruelling expedition in 1897 to the wild

Kimberley region in the state's far north, he was granted 5000 acres of prime pastoral land near what is now the town of Derby. Because he was also the local member for West Kimberley at the time, the Legislative Council's decision to grant the land to Alexander fuelled allegations of corruption. In 1897 he became lord mayor of Perth while continuing to sit in parliament.

With John entrenched as premier, critics of the Forrest family began to voice concerns that the state was being run by a clique. Allegations continued to emerge in the press that Alexander Forrest was engaged in shonky business dealings and had secured government contracts for his various companies. "The truth was that it would have been almost impossible for the government to have bypassed his many interests," wrote Alison and Dinee Muir. Geoffrey Bolton has adopted a cautious tone in analysing Alexander Forrest's fusion of money-making prowess and political influence. "As a capitalist he was an uncomplicated believer in the development of Western Australia's natural resources, enjoying the gamble of speculative investment, generous in prosperity and uninterested in the acquisition of power beyond the extent necessary to serve his immediate interests," Bolton wrote. "But he was careless of appearances and failed to realise that the casual practices of a small-town business community, linked by kinship and connexion, could be interpreted unkindly in a more sophisticated commercial and political milieu."

Alexander was dogged by rumours of corruption until his sudden death at fifty-one. He died with an estate worth £195,238, one of the largest amassed in the state at that time. A prominent statue of Alexander Forrest stands on St Georges Terrace in the heart of Perth's modern business district, an apt location for a memorial to an arch-capitalist.

After ten years as premier of Western Australia, John Forrest resigned in January 1901 to become the federal minister for defence in the newly formed conservative national government of Edmund Barton. At Australia's first federal election, in March of that year, he became the member for Swan. Forrest went on to serve as minister for home affairs, treasurer and acting prime minister in various conservative governments over the following seventeen years. But his greatest disappointment in politics came when he failed to win the leadership of the Liberal Party, then in opposition, by a single vote in 1913.

The opportunity arose after Alfred Deakin had resigned as leader, providing Forrest with the moment he had been waiting for since leaving the premier's office twelve years earlier. But Deakin, whom Forrest had regarded as a close ally and friend, threw his support behind Joseph Cook for the job and, five months later, the Liberals returned to power with Cook as prime minister. Deeply hurt by the probability that he had missed out on the prime ministership by one vote and saddened by what he felt was Deakin's betrayal, Forrest wrote a note to his old boss after the ballot: "My dear Deakin, 'Et tu, Brute?', Yours sincerely and in sorrow, John Forrest."

John Forrest achieved another of his great visions in 1917, when a 1700-kilometre transcontinental railway line was joined on the Nullarbor Plain, finally linking Perth to the eastern states. Sir John, who had championed the project for decades, saw it as vital to Australia's economic development. A year later, at the age of seventy-one, Forrest was seriously ill with a cancerous cyst on his temple. He resigned as federal treasurer and boarded a ship for London to seek specialist medical attention. But on 3 September 1918, with his ship off the coast of Sierra Leone, he died. When warned not to make the long journey to London, Forrest had said: "I have faced death before and I will face it now. What does it matter if I die at sea?"

To this day, Forrest's record of public service – his years as an explorer, his decade as premier of Western Australian and his eighteen years of prominence on the national stage – remains virtually unrivalled. His statue, a portly bronze figure swathed in robes, stands in Perth's Kings Park, overlooking the capital city of a state that he transformed.

2.

## MINDEROO

*I saw wealth on a scale which I found completely remarkable as a child. And I always wondered from that time on: how could you be involved in something like that?*
—Andrew Forrest

For Andrew Forrest, the thrills and spills of growing up at Minderoo in the 1960s were straight out of the pages of a Boys' Own adventure. If the scrawny blond kid wasn't swimming in the Ashburton River, which cuts through the vast pastoral station, he was dashing up and down its tree-lined banks trying to catch goannas and snakes – or perhaps a fish for lunch. Enclosed by the Pilbara's cavernous blue sky, Andrew and his two elder siblings, David and Janie, felt as if they were in the middle of nowhere. And they pretty much were. The nearest population centre to Minderoo was Onslow, a coastal town forty kilometres away with only a few hundred residents. The only real action was in Perth, but that was 1400 kilometres to the south and, even then, it was the world's most isolated capital city.

The Pilbara is an ancient land of stark redness that stretches forever, a dramatic terrain of spinifex plains broken by folded orange mountains and deep gorges formed over hundreds of millions of years. Coated in dust, Forrest would spend his days riding horses and playing in the bush. "We could get on a horse and ride for a day or two at a time in any direction and know we wouldn't run into anyone else," he recalled in 2005. Janie had similar memories: "You made your life with the animals and you made your life in the scrub and going out with the blackfella kids. It was great fun. You'd experiment with survival."

Until Forrest was eight, he was educated through the School of the Air, via a two-way radio in the Minderoo homestead, and at Onslow Primary School, where 80 per cent of the students were Aboriginal. When Andrew was old enough to attend boarding school, his mother, Judy, couldn't bring herself to send her youngest child all the way to Perth, so he went instead to an indigenous hostel at Carnarvon, about 500 kilometres south of Minderoo. It was here that Forrest learned to fight. A debilitating stutter made him a magnet for older bullies, and Forrest responded with his fists. But he also developed a respect for the indigenous students at the hostel. "These kids were smarter than me, they sang better than me, they could certainly fight better, and all round in sport, to use the words I can hear today, they licked me," he said in a 2009 speech. Forrest didn't last long at the Carnarvon hostel. "Mum pulled me out of the hostel because I wasn't receiving any of the letters she sent," he recalled. "The housemaster gave the mail to other kids to distribute. They often stole the letters looking for money, and once opened they would not be passed on."

Forrest's identity today is inescapably linked to his experiences growing up in the emptiness of the outback and the friendships he formed with Aboriginal people. As a boy he played with the children from the Thalanyji tribal group and was looked after by Thalanyji adults. "We grew up with a much more extended family that included dozens and dozens of indigenous kids and their mums and dads," he recalled. "We all grew up together on the back of an Australian stock horse. We were all taught how to track, how to hunt and to husband animals in the bush, both wild and bred …

The parents of my dozens of indigenous mates … mentored me as their own kids grew up with and around me."

One of these indigenous men, Scotty Black, had a particularly profound and lasting impact on Forrest. Black, the son of a Scottish man and an Aboriginal woman, came to work as a stockman at Minderoo as a 22-year-old not long after World War II. But he soon walked off the job, along with hundreds of other Aborigines across the Pilbara, as part of the 1946 indigenous stockmen's strike over their horrendously poor conditions. Black's half-brother was the legendary Aboriginal leader Peter Coppin, one of the main agitators behind the dispute. The pastoral strike, dubbed "Blackfellas' Eureka", was the first organised industrial action by Aborigines in the nation's history and is regarded as a key milestone in the battle for rights for indigenous people.

After three years of careful and secretive planning, the pastoral workers walked off the job on 1 May, just as the shearing season approached. They were protesting against laws that effectively made them the slaves of the white graziers. (In his modern and laudable crusade to rid the world of slavery, Andrew Forrest has not mentioned the indentured labour on which his own Minderoo property was built.) The station hands demanded freedom of movement and payment of thirty shillings a week – an amount still well below the minimum wage. At the time, most received only rations and the occasional clean shirt. In return for their labour, pastoralists had long granted Aborigines the right to permanent residence on the land. During the three-year strike – the longest in Australian history – dozens of Aboriginal strikers were thrown in jail and put in chains. The sheep and cattle stations of the Pilbara were paralysed without cheap labour and in 1949 some of the stockmen began to return to work with proper wages and improved conditions.

When Andrew's father, Don, took over the station in the 1950s, Scotty Black returned and went on to become the head stockman and a revered figure at Minderoo. The job of head stockman was one coveted by every white "ringer" who passed through. "I was in awe of him," Forrest said of Black in a 2010 interview with Perth's *Sunday Times* newspaper. "When we'd come back from Onslow Primary or the Carnarvon Hostel or boarding school, you'd leap into his arms and give him a hug and you'd get a huge hug back. We took orders from Scotty and we jumped to it without a problem. We just couldn't do enough for him. Scotty Black was my greatest mentor outside my own blood."

Black could ride like the wind and, as the assistant manager of the station, had responsibility for thirty employees whenever Don was away. Yet he still lived a traditional Aboriginal life as a senior law man, performing initiation ceremonies, circumcisions and corroborees. In the impressionable eyes of the young Andrew Forrest, Black's ability to walk in two worlds was proof that Aborigines can balance the professional demands of a white man's world with their ancient traditions – a belief that would later form the foundation of his thinking on indigenous people.

As Forrest has used his wealth and influence to campaign to end the shame of indigenous disadvantage in recent years, he has repeatedly invoked the name of the late Scotty Black, who has been elevated in death to the status of outback hero. One family anecdote that has taken on almost mythical proportions is the day Forrest went missing on the station as a toddler, clad only in a towelling nappy. After slipping through a gate and wandering away from the homestead, Andrew fell through a cattle grid and was trapped as heavy vehicles passed overhead and the blazing Pilbara sun beat down on his pale skin. A search party could find no sign of the boy. Sure enough, it was Scotty Black who discovered the terrified two-year-old under the cattle grid and pulled him to safety. From then on, Black became a second father to Forrest. The boy would sneak into the stockman's room while he was out working, lie on his bed and wait for him to return.

Several years later, Forrest was on a cattle muster with Black when they came across a bull that had fallen into a gully and become trapped. Faced with the choice of leaving the bull for two days to return home for an axe to cut the tree roots which had entangled the beast, or trying to free it immediately, Forrest put his whip-cracking skills into action and drove the animal out. Black was impressed with the boy's actions. "He told me: 'Young fella, you've just done a bloody man's job – you can go places, son.' And I've never forgotten. He gave me belief in myself." To this day, one of Forrest's most treasured possessions is a photograph of himself with Scotty Black taken on the banks of Ashburton River when he was about nine years old. The pale, skinny kid, wearing only his swimming trunks, is holding tree branches and performing a mock corroboree with the smiling Black on the parched red earth of Minderoo. The photograph, taken by Don Forrest, shows Black as a tall, bespectacled and fit-looking man in his mid forties. There is nothing in the vast background but a distant haze on the horizon.

Ever since Andrew's great-grandfather David Forrest established Minderoo in the 1870s, the family has shared the property with the Thalanyji people, who have continued hunting and performing tribal customs at their sacred sites that dot the 230,000-hectare property. David Forrest was born in 1852 near Bunbury and followed his four older brothers – including John and Alexander – to Bishop Hale's school in Perth. When the Forrest brothers secured a lease of Pilbara land, they chose David, who had been working with sheep in the Wheatbelt region in the south of the state, to run it. In 1874, the plucky 21-year-old set out with four Aborigines from the town of Quairading, east of Perth, with 2000 sheep for an epic 1600-kilometre journey that would take them five months.

Towards the end of the gruelling trip, as he pushed into Thalanyji country near the Ashburton River, Forrest was speared by an Aborigine who had accompanied him on the journey for several days. The motive for the attack is not entirely clear, but the official Forrest family version is that tensions erupted when the Thalanyji, who by that time had befriended David Forrest, ordered the unknown Aborigine to stay out of their country. In retaliation, the "wild native" drove his spear into Forrest's shoulder one night as he slept on the ground. Almost unconscious, and with the spear protruding from his shoulder, Forrest killed the attacker with his shotgun and was nursed by the Thalanyji people. When the party reached Roebourne in May 1878, David told the magistrate he had shot the man in self-defence. For Andrew Forrest, this story has been central to his understanding of his family's relationship with the land and its traditional owners for four generations: "You kind of felt a real at-oneness with the land and with the [Thalanyji] people, and if they hadn't saved him, there'd be no Forrests."

David Forrest returned to Perth to recuperate in hospital for several months before making a second trek to Minderoo with another 2000 sheep. A wild storm later destroyed most of the station's sheep, along with many of the improvements he had made to the station. But David, still in his twenties, made a fresh start, moving the homestead to a more protected site about forty kilometres south of Onslow, where it stands today.

Life at Minderoo in the 1880s was harsh and often miserable. David's wife, Mary, was the only white woman in the district and her nearest neighbour was almost 250 kilometres away in Roebourne. The couple's first two sons, John and William, both died as infants at Minderoo and were buried beside each other in the red dirt. A third son, David, was born on the remote station in 1889 but died in Perth later that year. The surviving children were two boys, Mervyn and Don, and two girls, Lily and Violet. Don died in World War I, leaving Merv – Andrew Forrest's grandfather – as David Forrest's only remaining son and the sole heir to Minderoo.

Under Merv's strict management, Minderoo was transformed into a thriving sheep station, with several new buildings, extensive fencing, forty windmills and seventeen dams added to the property. By the 1930s the station carried more than 50,000 sheep and produced an annual wool clip of more than 1000 bales. Years later, at his grandfather's knee, the young Andrew Forrest was told of the importance of Minderoo and the Forrest dynasty. "My strongest memory of Grandad," Forrest said in 2005, "was going into his study and being shown the enormous photograph albums of the opening of the Kalgoorlie water pipeline and the other great endeavours that his uncle and his father had done."

As the years went on, Merv became as hard as the Pilbara earth. "I don't think Dad enjoyed it. I think he found it a hell of a life. He seemed to lose all his gaiety which I know he had as a younger man," recalled his daughter Shirley. Merv was particularly cruel to his son Don. "He used to tell me what to do, he'd never ask me what I'd like to do," Don said. "I would have liked to have gone to university. 'No, you don't want to go to university, son. Your life's on the land.' He wouldn't take time with children to explain. That was a bit sad; we never had a friendly relationship, ever." Shirley Forrest concurred: "Dad was awful to Don, but quite frankly I never felt like tackling my father. His word was law. It was like talking about God."

On his father's orders, Don Forrest became the manager of Minderoo in 1951, when he was just twenty-three. As he toiled in the Pilbara heat, Don felt the burden of his pedigree. "When it was stinking hot the sun'd be on your back and the flies would be worrying you and you'd think, 'What the hell am I doing in this

God-awful country for?'" Don said. "But my grandfather was the pioneer, my father was the developer and I had to just keep it all going." His wife, Judy, recalled the menacing dust storms that would roll in almost every afternoon. "I could cope with the heat, but the dust storms: you couldn't put your hand on anything without it coming up red," she said. "You couldn't wipe your arm across a perfectly clean table." Don and Judy bought the property outright in 1974, a year before Merv died in Perth at the age of eighty-four. Their three children – David, Janie and Andrew – would be the only direct descendants of David Forrest to retain the famous name. Andrew, in particular, bore a striking resemblance to his great-grandfather as he grew.

Don Forrest had met Judy during a holiday in Perth a few years after taking over Minderoo. The young city-dweller was swept off her feet and soon they were married and living at Minderoo. "I don't know what I went into," Judy recalled. "I just thought Don was the greatest thing on earth so it didn't matter where we went. In fact, as we drove over the boundary fence for the first time after we were married, I burst into tears because suddenly I was there, and I thought, 'Wow! What have I taken on?'"

Like his own father, Don Forrest was a strict parent and a strong-willed man. But with his often rebellious and tenacious youngest son, he had every reason to be firm. When Andrew was about ten, Don forced him to chew a mouthful of soap after he was caught swearing at his sister while saddling up horses. Another time, Andrew was secretly playing with matches and petrol, when he badly singed himself, including his face.

Regardless of Andrew's misbehaviour, Don Forrest was a hard man to please. Years later, even when Andrew Forrest had built Fortescue Metals Group into a mining powerhouse, Don was reluctant to tell his son he was proud of him. Forrest attributed his father's stern approach to parenting to his own upbringing and contended that it was his strong mother who instilled in him a sense of self-belief.

Outside the boundaries of Minderoo, the landscape of the Pilbara was being altered irrevocably during Forrest's childhood. In November 1952, an eccentric Pilbara pastoralist and prospector called Lang Hancock had been flying his plane low, in blinding rain, through the Hamersley Ranges, about 200 kilometres inland from Minderoo, when he glanced down at the glistening rust-red rocks below. Suspecting the presence of iron ore, Hancock made a mental note of the location and later returned to take samples. He soon came to realise he had discovered the world's largest deposit of iron ore – more than 1 billion tonnes of the mineral.

But instead of trumpeting the find, Hancock and his business partner, Peter Wright, kept it a secret for almost ten years. At the time, the federal government had an embargo on exporting iron ore due to a mistaken perception that the nation's reserves were scarce, and Hancock was therefore unable to peg the tenements. He set about lobbying the government to lift the ban, which finally occurred in 1960. In truth, it had been known in geological circles since the late 1800s that it was likely the Pilbara contained a bounty of undiscovered iron ore. But it was only after World War II that the high cost of mining and shipping the commodity – the key ingredient in modern steelmaking – could be justified by demand emanating from the booming economies of Japan, Europe and North America.

After furiously lobbying global companies to back them, Hancock and Wright eventually struck an eye-poppingly lucrative deal with London-based miner Rio Tinto, which handed them a 2.5 per cent royalty in perpetuity on every tonne of iron ore exported from the Hamersley Ranges. Six decades later, that agreement would help propel Lang Hancock's daughter, Gina Rinehart, to the title of Australia's richest person, overtaking Andrew Forrest on the *BRW* Rich List. Soon after the Rio Tinto deal was sealed in the early 1960s, new towns began springing up across the Pilbara as global capital poured in. The Pilbara's population soared tenfold and by August 1966 the first shipment of 52,000 tonnes of iron ore had left the newly built harbour at Dampier, bound for the steel mills of rapidly industrialising Japan.

The pace of construction was astonishing. It had been just nineteen months since Hamersley Iron – a joint venture between Rio Tinto and the US Kaiser Steel Corporation – had begun creating towns at Tom Price and Dampier while building an open-cut mine, 320 kilometres of heavy-duty railway, two power stations and a deepwater port. By 1968, American company Bechtel Pacific had started producing iron ore at Mount Whaleback, in a lonely stretch of desert 400 kilometres from the coast in the eastern Pilbara. Soon taken over by mining giant BHP, Mount Whaleback now ranks as the world's biggest open-cut iron ore mine, a giant crater measuring five kilometres long, 1.5 kilometres wide and 430 metres deep.

The arrival of large-scale mining in the Pilbara inevitably caught the attention of the young Andrew Forrest, who had a glimpse of his future one day while passing the Cape Lambert iron ore plant near Karratha, about 200 kilometres north-east of Onslow. "We came across industry of sight and sound I could never have imagined," Forrest said in 2005. "And I just thought: what is this doing here? I thought the whole country revolved around sheep and cattle and hard yakka. I saw wealth on a scale which I found completely remarkable as a child. And I always wondered from that time on: how could you be involved in something like that?" Forrest would return to the Cape Lambert project as a seventeen-year-old to gain his first taste of the mining industry, doing summer work as a process engineer's assistant.

The rapid industrialisation of the Pilbara in the late 1960s created thousands of jobs for Australians and enriched shareholders and governments. But it did nothing for the original inhabitants of the ancient landscape. Aboriginal people have lived in the Pilbara for at least 30,000 years and have a powerful, spiritual relationship with the land. Only in the past 140 years have they had to share the rugged country with white Australians. Yet when the deals between the state government of Sir David Brand and the international mining houses were drawn up to unearth the Pilbara's bounty of iron ore, Aboriginal people were not even invited into the room. Not a cent in royalties was paid to indigenous people, and the mining companies preferred to fly in labour from thousands of kilometres away rather than train local Aborigines. By the early 1980s, more than a decade after it opened, the huge Mount Whaleback mine had engaged just fifteen Aborigines. Whereas Aborigines had been long relied upon by pastoralists for their bush skills, they were virtually ignored by the emerging mining industry.

In 1960s Australia, however, attitudes towards Aborigines in some quarters were beginning to change. The 1967 referendum that sought to remove clauses discriminating against Aborigines received an overwhelming 90 per cent "yes" vote across the nation. (Western Australia had the lowest percentage among the states, with an endorsement of 79 per cent – and this was much lower in most rural areas.) A year earlier, the Federal Court had ruled that indigenous station hands should receive equal pay and conditions to their white counterparts – a decision that would have pleased Scotty Black and his fellow stockmen at Minderoo. Many pastoralists, however, were opposed to the judgment and sacked their Aboriginal workers. Historian Geoffrey Bolton summarised the devastating impact of these sackings, which helped create the conditions for decades of social problems that have never been addressed in the Pilbara. "With little demand for their skills, limited job prospects, and legal access to alcohol," he wrote, "these displaced Aboriginal communities faced daunting social problems beyond the imaginative grasp of most Western Australians."

In Perth, David Brand and his zealous minister for industrial development, Charles Court, remained staunchly opposed to any form of Aboriginal land rights even as the Pilbara mining industry flourished throughout the 1960s. Court's hardline views were best illuminated in the infamous Noonkanbah dispute when he was premier in 1980 – an incident that marked a new low in the government's relations with Aboriginal people. An American drilling company, Amax, had applied to explore for oil at Noonkanbah pastoral station in the Kimberley region, to the north of the Pilbara. But the owners of the land, the Yungngora people, complained that their sacred sites would be disturbed and Amax was reluctant to enter without the landowners' approval. Court, as premier, ended the stand-off by ordering a police-escorted convoy of forty-five drilling rigs onto Noonkanbah, leading to violent clashes and noisy protests. His brazen approach was lauded by many, reflecting the prejudice prevalent in WA mining circles, and society generally, at the time. One avid supporter of Court's stance was Lang Hancock, the father of the Pilbara iron ore industry, who appeared on television in the early 1980s to offer his views on how to solve "the Aboriginal problem". Hancock told *Today Tonight* he believed "no-good half-caste" Aborigines should be herded into one area and their waterholes poisoned so they could "breed themselves out".

As a huge new industry sprung up around him in the Pilbara scrub, and while his Aboriginal mates continued to spend their days catching barramundi in the Ashburton River, the eight-year-old Andrew Forrest flew out of Minderoo bound for one of Perth's wealthiest boarding schools. Judy believes that sending all three of her children away from home at such young ages was the most emotionally wrenching part of station life. "It was always hateful, absolutely dreadful," she recalled. "We'd all be glassy-eyed."

Andrew's elder brother, David, recalled the children's pain of returning to Perth after school holidays spent at Minderoo: "The last day, you'd try and get as much red dust in your hand and smear it all over you. And you'd have to get it all washed off, of course, because you'd be made to clean up to hop on the plane. But at least you could get it ingrained in your hand."

As he flew out to Perth one day in the early 1970s, his pores full of red dirt, Andrew Forrest could have had no idea how far his life's journey would take him.

# 3.

# FISTICUFFS AND FAILURES

*The wildest colts often make the best horses.*
—Ken Tregonning, former Hale School headmaster

With its sandstone buildings, lush lawns and Swan River frontage, Christ Church Grammar is the school of choice for Western Australia's wealthy. It is Perth's equivalent of Melbourne Grammar or Sydney's Cranbrook, churning out educated young men with a network of ready-made connections. And like those institutions, Christ Church tends to boast of the triumphs of its alumni. Among those to have worn the blue-and-gold school tie are former British Airways chief executive Rod Eddington, Western Australia's chief justice Wayne Martin, national men's hockey coach Ric Charlesworth, entrepreneur Peter Holmes à Court, renowned solo yachtsman Jon Sanders, comedian Tim Minchin and the lead singer of rock band The Triffids, David McComb, who died in 1999.

Christ Church also likes to claim Andrew Forrest as one of its "old boys", referring to him in school literature as a member of the class of 1978. In truth, however, Forrest never made it that far, and many at Christ Church were relieved to see the back of him when he left abruptly in 1977.

Forrest arrived at Christ Church, in Perth's old-money suburb of Claremont, in 1970 as an eight-year-old with a severe stutter that badly dented his self-confidence. "I always had the stutter, I grew up with it," he later recalled. In fact, Forrest had inherited the disorder from his father Don. "I'm amazed he speaks so well now [because] he used to stammer at school and at home," Don Forrest said in 2007. "I was a stammerer and Andrew took it up."

Forrest was placed in Walters boarding house, which fellow students of the era remember as a Dickensian place. There were ten to fourteen beds to a room, the mattresses were thin, the food was terrible and the spartan dormitories were icy-cold in winter. The bathrooms in the boarding house consisted of several showers in a row, with no partitions and a highly unreliable supply of hot water. By the late 1970s some effort was being made to redress these privations. The school yearbook of 1977 – when Forrest was in Year 11 – records that the boarders' common room had just been fitted with "carpet, curtains, heating and colourful posters, which all helped to make the room much more pleasant". Christ Church may have been expensive and prestigious, but clearly the boarders were not living in luxury.

Forrest's boarding master at Walters house, Tony London, was a key figure among the staff at the school. He recalls that bullying was rife when he took over the boarding house, likening the environment to the savagery in William Golding's classic *Lord of the Flies*. Younger pupils, he says, were forced to be subservient to the senior boys, mainly the Year 12s, who had the power to decide how much the smaller boys could eat and to dish out tasks and punishment as they saw fit. Along with other senior staff, London introduced key reforms during Forrest's time at Walters, including a smorgasbord which prevented the Year 12s from controlling the food. "Walters house had been run very loosely; the boys had the run of the house," he says.

The long-serving headmaster at Christ Church was Peter Moyes, a conservative educator who had won plenty of kudos for reviving the school's fortunes. But, as one ex-student notes, Moyes was "well past his use-by date by the 1970s". By all accounts, there was little room for freedom of expression and the cane was employed frequently to deal with all forms of insubordination. If boarders left the school grounds on a Sunday, it was only after attending the compulsory 7am church service, and they were expected to wear the full uniform, including the blazer and tie.

One legendary sports teacher, Akos Kovacs, who died in 2012, was known for frequently striking students with his bare hands. During swimming lessons in the Swan River, Kovacs would use a long pole to push boys into the water. Only if a hapless youth was close to drowning would Kovacs, an eastern European immigrant and World War II veteran, bring out the pole at the last second to save him.

Forrest's brother, David, who is two years older, was ensconced in this harsh environment at Christ Church when Andrew arrived fresh from the Pilbara. David's widely used nickname at school was "Twig", a play on the family name, but it would be several years before Andrew also came to be known by that moniker. In fact, he appears to have inherited it only after David left school; for Andrew, "Twig" may also have been a reference to his exceptionally scrawny physique as a youth. In the school annual of 1977, history of sorts was made when the name "Twiggy Forrest" was used for the first time in relation to Andrew. It would be thirty years before it became the best known nickname in corporate Australia.

People recall Forrest as difficult and disruptive in class, rude to teachers and disliked by many of the boys at Christ Church. A more generous view is that he was a larrikin in a hostile environment who was deeply anxious and insecure because of his stutter and desperately sad at being more than 1000 kilometres away from home. A former classmate, Rohan Pixley, says Forrest loved the land and was always talking to the other boys about his fondness for Minderoo. "Twiggy hated coming back to boarding school," he says. "He would have wanted to be back up on the property." One former student recalls Forrest taunting a fellow student who had a large nose by calling him "Arrowhead". "He used to put his hands over the top of his head to make an arrow shape every time the poor kid walked past him," he says.

David Forrest has claimed publicly that Andrew was bullied by students and victimised by school authorities, referring to a "boomerang elbow" his brother still carried as a result of being pushed down the stairs by a teacher. But it's difficult to find anyone who can confirm that incident and David declined to be interviewed, saying Andrew would not allow him to speak. A number of those who were there at the time say Forrest instigated much of the violence in which he became involved. One former student, who is now a leading businessman in Perth and wants to remain anonymous, recalls: "He was very quick with his fists. He claimed to have learned to fight from the Aboriginal kids on Minderoo station and he would fight in a way that put an end to the argument very quickly – a really hard punch to the nose or the eye, or both."

Forrest wasn't particularly popular at Christ Church and he was certainly not academically bright. He was enrolled in the lowest level of mathematics, known as Maths IV – proof that the future master of financial markets was no whiz with numbers as a youth. Forrest's main claim to fame around Christ Church appears to have been as a keen, and particularly aggressive, member of the school rugby team. He was also a decent swimmer from the moment he arrived. A photograph of the prep school swimming team from the 1973 school annual shows Forrest as a rake-thin eleven-year-old with pale skin and a mop of sandy hair. Sitting casually with his arms crossed, he is almost reclining in his seat while the other boys make an effort to sit up straight.

David Forrest told journalist Mark Drummond in 2007 that one teacher at Christ Church had written on Andrew's school report that it was lucky he had a sheep and cattle station to go back to because he wouldn't amount to much in life. Over time it has become accepted – by Forrest and many others – that Tony London was the author of that remark. But London, who was an English teacher as well as Forrest's boarding master, rejects this, saying he would never have written such a remark and headmaster Peter Moyes would never have allowed it on any student's report.

London has also heard talk over the years that Forrest blames him for his deep unhappiness at Christ Church. "He was a very unhappy boy – I felt sorry for him," says London, who ran into Forrest in 2007 and finally made peace with him, telling him he had achieved much more than anyone had expected. Like many others at Christ Church at the time, London is adamant that Forrest brought on much of the bullying he received. "Andrew divided people one way or the other and he would say things which would stir people up," he says. "He said he was bullied and I believe that was the case. Unfortunately, people who are bullied sometimes attract it."

Then one day towards the end of the 1977 school year, Forrest was gone. It's not entirely clear whether he was asked to leave Christ Church or whether his parents decided he'd been in enough trouble and needed a fresh start. Either way, his abrupt departure appears to have been precipitated by his decision to pick a fight with a bigger student called John Weatherhead. "He was fighting one morning and he was gone from school that afternoon," says one student who was in Forrest's year at Christ Church.

Weatherhead says he clearly recalls the events of that day – and he still sports a small scar above his right eye as a reminder of the young Twiggy's punching ability. "I was standing at the top of the stairs when Forrest, who was down below with a group of mates, looked up at and yelled out, 'Poofter!' I was quite shocked because I'd never even met the guy. And then it escalated. I said, 'Come here and say that,' and he said 'Poofter!' again. There was some pushing and shoving and then punches started flying. I copped one in the eye, which needed a stitch, but the only other recollection I have of the fight is I had his blazer over his head and I had him on the edge of the steps. I was in two minds – I thought, I could just push him down here. It was a split-second thing. I thought, I will hurt the guy too bad. That's when he came up with a blind left hook, because he had his jacket over his head. Then it got broken up by teachers."

Weatherhead says he bears no malice towards Forrest and the pair have been friendly towards each other during a few chance encounters around Perth since school days. "After our encounter, he always gave me the impression he felt somewhat remorseful and wanted to make amends, even years later when we ran into each other," he says.

Another former student, Bruce "Jock" Strapp, has a different recollection of the brawl that appears to have been the final straw for Forrest's teachers and parents. Strapp says he was walking down the stairs with Forrest when Weatherhead made a joke about Forrest's stutter. In itself, that was unsurprising to Strapp because students would regularly make fun of Forrest.

After his abrupt departure from Christ Church towards the end of Year 11, Forrest was enrolled at Hale School the following year to fulfil what must have been his parents' fervent wish for their wayward son to complete his secondary education. One of the state's oldest and most illustrious schools, Hale is set over forty-eight hectares in Wembley Downs, one of the many comfortable and serene suburbs that lie between Perth's CBD and the sapphire waters of the Indian Ocean. Like most private boys' schools, Hale has an integral focus on sport. These days, the campus includes an Olympic-sized heated swimming pool, sixteen tennis courts, four football fields, five cricket ovals, four soccer pitches and a gymnasium with basketball, badminton, volleyball, squash and rock-climbing facilities.

Hale's list of alumni is possibly even more impressive than Christ Church's. The school has produced six premiers of Western Australia (including Sir John Forrest), thirteen Rhodes scholars and a bevy of prominent musicians, artists, scientists and captains of industry. Hale is nowhere near as blue-blooded as Christ Church; its location in Wembley Downs, slightly north of Perth's wealthiest enclaves, means it draws more students from the city's middle-class suburbs.

When Andrew Forrest arrived at Hale as a boarder in early 1978, with a reputation for pugilism and poor marks, he would have been viewed as the one of the least likely candidates to follow in the footsteps of the school's famous graduates. Forrest appears to have been extraordinarily lucky that Hale took him in at all. The school's former long-serving headmaster, Ken Tregonning, reveals that Forrest's father, Don, an old friend and a Hale alumnus, approached him personally to ask whether he could enrol his son to repeat Year 11. He admits he didn't ask Don too many questions about why Andrew had left Christ Church. "I didn't want to pry," he says. "I said 'yes' because he [Andrew] was the son of an old boy and I was good friends with Don."

Tregonning, a keen historian, was well aware that the Forrest clan had a long association with Hale, dating back to the education of John, Alexander and David Forrest. These days, Twiggy's only son, Sydney, is a student at Hale and Forrest is recognised as a generous donor to the school. In 2009 he was invited to open the Forrest Library, which was named in recognition of the family's association with Hale and was the beneficiary of a fat cheque from Forrest. That same year, the school magazine lauded Forrest alongside fellow old boys such as Wesfarmers chief executive Richard Goyder and former WA Liberal treasurer Christian Porter as examples of alumni who had "seized the day" and achieved great things. "They are the sorts of people who haven't been too tired when things were a bit hard and they definitely didn't put things off until tomorrow," the magazine said.

The move to Hale was the seminal event in the young Andrew Forrest's life. "There was no way if he stayed at Christ Church that he was going to pass Year 12," says former classmate Rohan Pixley. "Hale deserves a lot of credit for that." Tregonning, as headmaster, appears to have played a central role in the transformation of the sixteen-year-old Forrest. "The wildest colts often make the best horses," says Tregonning, now comfortably retired in the posh Perth suburb of Peppermint Grove. "I used to relish boys who were in trouble. Often all they need is encouragement, guidance and praise, and they will come good."

One morning quite early in Forrest's time at Hale, Tregonning decided to stroll across to the school pool before breakfast to watch the swimming squad train for the upcoming Public Schools Association carnival, a prestigious event on the calendar of Perth's private boys' schools. But Forrest hadn't turned up to the early-morning training session, prompting an angry headmaster to storm over to the boarding house and wake the youth from his slumber. For Forrest, the image of an irate white-haired headmaster in his dressing gown at the foot of his bed at 6am seems to have been a proverbial light-bulb moment. "He told me long after that it was the first time somebody had ever really wanted him," Tregonning recalls.

The official line is that Forrest became a model student under this sort of guidance at Hale School. He did become a school prefect and a boarding house captain in his final year and made the rugby and swimming teams. "Hale School represented a turning point in my life," Forrest wrote in a message published in a 2012 newsletter for the Hale School Foundation. "To attend a school which believed in me and my potential allowed me to achieve well above what I ever had in many areas. On reflection, the support and drive that was instilled in me can be traced back to the staff, many of whom saw more in me than I did myself."

The truth about Forrest's turnaround, as with many events in his colourful life, seems to be a bit less clear-cut. Former students recall that Forrest never lost his larrikin streak at Hale and was often the leader of the pack whenever the boys were getting up to mischief. But many also recognised at the time that their mate was likely destined for bigger things. "You knew if you were going out somewhere with Andrew you'd have some fun and that he'd end up pushing the envelope a bit," says one former student who is still close to Forrest. "He was great fun and larger than life. Even then, it was obvious that his success would come from a combination of personality, hard work, the gift of the gab and his contacts."

Forrest also brought to Hale his love of a good scrap. Some recall him instigating a brawl with students from Churchlands Senior High School, the local government-run school, outside a nearby shopping centre. A tale still circulating among Forrest's old classmates is that he would always carry a mouthguard in his back pocket just in case the opportunity arose for a fight.

During his two years at Hale, Forrest claims to have suddenly lost his stutter, which had always grown noticeably worse whenever he was under pressure or before an audience. The teenager's parents had earlier sent him to a hypnotist, who had promised to cure the stutter with just three expensive lessons. "After about the ninth or tenth lesson, we realised he was a snake oil salesman and we stopped," he recalled. According to Forrest, he suddenly "lost" the stutter by volunteering for the debating team. It was a valuable lesson to the youthful Twiggy that risk could bring great rewards: "It was a real blood sport at school, a gladiatorial environment – all jeering and hissing and booing. I was the first speaker – the weak link – and the auditorium was packed because they all knew I had a stutter. I just cracked it. I managed to speak slowly and softly. I knew I could do it and I have never looked back. That taught me the power of real determination."

There is no complete cure for stammering, and it's well documented that many successful people – from King George VI to Hollywood actors and politicians – have learned to control the disorder to the point where it is almost undetectable. Forrest has never spoken in depth about the psychological impact of the stutter on him as both a child and an adult. He suffered from it for many years after leaving Hale and even into his late twenties and early thirties. "His stutter was still quite pronounced, even in Year 11 and 12," confirms an old friend. "He was very conscious of it and worked on it. Being on the debating team did help him and it gave him confidence. But it certainly hadn't gone by the time he finished school."

Forrest's confidence was also rising rapidly when it came to chatting up girls. "He was very persistent in asking girls out," recalls the old friend. "He could pull girls like no one I knew because he just didn't give up – and he did have a charm."

His academic performance appears to have lifted at Hale, to the point where Forrest had the marks from his final exams to enrol in a commerce degree at the University of Western Australia, the state's most prestigious tertiary institution. The marks needed to enrol in commerce were quite high, but still well below those required for courses such as law and engineering. Tregonning believes the fact that Forrest repeated Year 11 at Hale in 1978 helped him compete academically with boys who were a year younger than him and also gave him a chance to mature. But students from that period say it would be wrong to suggest that Forrest was one of the brighter pupils in the class of 1979. "He was reasonably good academically," says a classmate. "But I don't think he was an outstanding student academically. He did well enough to get into university but he certainly wouldn't have been in the top ten guys in our school."

Forrest's close mate at Hale, Adam Rankine-Wilson, was another who never excelled academically. Rankine-Wilson, who died of cancer, aged forty-nine, in 2011, started his career as an insurance salesman with AMP but went on make a fortune in investment banking and was not afraid to show off his money-making ability by acquiring flashy toys. Both Forrest and Rankine-Wilson had qualities that can't always be taught in school: street smarts and self-belief in spades. Forrest may have been reflecting on himself and Rankine-Wilson in 2010 when he said: "The world is full of intellectuals that failed and ordinary people that you would never have rated at school, that are successful. The difference between them is determination." For their thirty-year Hale anniversary reunion in 2009, Forrest and Rankine-Wilson, the billionaire miner and the deal-making whiz, got the chance to show off their financial success by taking their old schoolmates – many of whom were still paying off mortgages and living ordinary lives – to Rottnest Island on their huge launches.

One of Forrest's teachers at Hale, Bill Edgar, remembers well the teenaged "Twig" and says he is never particularly surprised when boarders from yesteryear make good. Edgar, who is these days the official archivist at Hale, believes boarding schools breed character and toughness in young men, and that this may well be the secret to much of Forrest's resilience in later life.

After school, Forrest sated his fledgling appetite for adventure by flying to Africa with his Hale classmate John Morrison, who these days is a top investment banker at Grant Samuel in Melbourne and remains one of Twiggy's closest friends. The pair headed straight for the Ugandan capital, Kampala, where Forrest bluffed his way into a meeting with the country's tourism minister on the basis that he was there exploring investment opportunities. The youthful Forrest, showing all the talents that would later land him with the nickname "Silver Tongue" in Perth stockbroking circles, managed to persuade the Ugandan minister to give him a special certificate which guaranteed the two Australian visitors unimpeded access to the country's best tourist sights.

But while Forrest and Morrison were snorkelling on a pristine coral reef about twenty kilometres offshore, a huge wave sank the small fishing boat they had been travelling in. Twiggy reportedly helped those on the boat, including a mother and her baby, to scamper onto a rock, where they all stayed for two days without food or drinking water. Finally, the bedraggled group, by this time close to dehydration, was rescued by a passing boat. The African adventure story has been told to a few journalists over the years, although we may never know the precise events that took place or the extent of the duo's bravery. Regardless of the facts, it wasn't to be the last time that Twiggy's smooth talking and penchant for risk would lead to unexpected problems.

Back in Perth, Forrest started his commerce degree but soon switched to UWA's arts faculty, completing a double major in economics and international politics. He then returned to Minderoo in 1982 to ponder his career options. By now, his parents' marriage had ended and life on Minderoo, where droughts, floods and cyclones were frequent events, was becoming even tougher for Don Forrest, who was in his mid fifties. (Don later remarried and stayed another sixteen years at Minderoo, before selling the station in 1998 to pay his mounting debts and moving to Perth.)

If Andrew had any lingering thoughts about whether he might remain on the land and take over the family business, his father quickly dispelled them. "I'd kind of been given the impression by Dad that Minderoo would most likely go on a hands-on basis to my older brother David and that was fine by me," Forrest said.

"And if I had a role it would be of a more investor/accountant/financial advisory role. So reading the tea leaves it was like, 'You'd better go and make your own way, son,' and I think that was probably a very good thing."

According to Don Forrest, it was Andrew who showed the least interest of the three children in staying at Minderoo. "He was never really interested in cattle and sheep; he was much more interested in the business side of things," he said.

Forrest spent several months working for his father at Minderoo before heading back to Perth in search of a job. It was 1983, Perth was gearing up for yet another economic boom and Forrest was about to discover his first true calling in life: making money.

## 4.

## THE WILD WEST

*He's got a hide on him. You could shoot him
in the arse and he wouldn't know.*
—JEFF BRAYSICH, Forrest's former stockbroking colleague

John Poynton looked across his desk at the stammering country bumpkin sitting opposite him and tried to suppress a chuckle. Poynton was a star stockbroker and partner at Hartley Poynton, the blue-blooded firm his father had established in Perth during the 1950s. The unemployed Andrew Forrest, who was "dressed like a stockman" and armed with only an economics degree and a healthy dose of chutzpah, had managed to sweet-talk his way into Poynton's office.

But when it came to speaking face-to-face, Forrest's chronic stutter meant he could barely get his words out. "It was hard to be serious because I didn't know how he would ever be able to sell anything to anybody – clients were going to hang up before he'd got the pitch out," Poynton recalls of the 1983 job interview. "I'm assuming that, like a lot of stutterers, the more nervous he was the worse it became. But this was so bad it was almost like a caricature of a stutterer." Poynton sent Forrest on his way. Thirty years after that meeting, he sees irony in the fact that Forrest – who is now a friend – would ultimately make billions of dollars through his gift of the gab. "I'm the idiot for not employing him," he laughs.

Dressed in RM Williams boots, moleskin chinos and a blue chambray shirt, the tousle-haired Forrest did not exactly fit the image of the employee favoured by Perth's clutch of establishment broking firms. They preferred to hire young men in dark suits raised in the city's genteel western suburbs. Day after day, Forrest trudged along St Georges Terrace looking for a break. He cold-called and sent letters to most of Perth's broking firms and banks, but he continued to be rejected.

Eventually, he promised Bruce Benney, the head of broking house Benney Partners, that he'd work for no pay for if he was just allowed in the door. The offer was too good to refuse for Benney, who took on Forrest as a research analyst. "He was a terribly, terribly enthusiastic young man," Benney recalled. "It was through sheer persistence that he got his foot in the door. And once he did, we couldn't close it." After three months, Forrest was put on the books as an employee and back-paid. The real adventure had begun.

Forrest's arrival in the stockbroking industry in 1983 proved to be fortuitous. It was the beginning of a gilded four years of wealth accumulation and corporate deal-making in Australia that would end abruptly with the October 1987 stockmarket crash. By then, plenty of fortunes and reputations would have been made and lost. And Perth would produce more than its fair share of high flyers, several of whom would end up in jail.

The global economy in 1983 was just recovering from the effects of a sharp recession in the United States, while in Australia the newly elected Hawke government was preparing to launch a string of business-friendly reforms that would open up the economy to the rest of the world. These included floating the dollar, deregulating the financial system and slashing tariffs. In Perth, with its history of money fever dating from the nineteenth-century gold rush, the boom of the 1980s would be bigger than anywhere else in the country. So much bigger, in fact, that Western Australia would officially declare itself the "State of Excitement"; the slogan was stamped on every new vehicle's licence plate.

Yet it was a yachting race off the coast of Rhode Island in 1983 that had an even bigger impact on the Western Australian psyche. In September of that year, just as Forrest was settling into his new job at Benneys, a highly exuberant Perth entrepreneur called Alan Bond spearheaded Australia's bid to wrest the America's Cup from the venerable New York Yacht Club. Bond was already a major figure in Australian business but his bankrolling of this national sporting conquest transformed him into a popular hero and propelled him towards bigger and riskier deals that would lead him into bankruptcy, and ultimately prison for committing Australia's biggest corporate fraud. Bob Hawke, the former Perth beer-drinking champion turned prime minister and a big fan of Bond, summed up the ebullience of the America's Cup triumph with the immortal words: "Any boss who sacks anyone for not turning up today is a bum!"

When the Auld Mug arrived in Perth after 126 years at the New York Yacht Club, an estimated 200,000 people – one-fifth of the city's population at the time – turned out to catch a glimpse of a beaming Alan Bond in an open-top Rolls-Royce leading the ticker-tape parade. Spurred by his victory, "Bondy" went on to acquire all the trappings of wealth – a private jet, fast boats, grand properties, prized artworks and blonde mistresses – while doing multibillion-dollar business deals at a dizzying pace, all funded entirely with other people's money. At the peak of his hubris, he even starred in a popular television advertisement for his own Swan Brewery featuring a catchy jingle with the chorus, "They said you'd never make it, but you finally came through." Bond had started to believe his own legend.

One of Bond's closest mates was Laurie Connell, a pugnacious Perth business identity whose twin passions in life were doing deals and betting colossal amounts on horseracing. Known along St Georges Terrace as "Last Resort Laurie", Connell could always be relied upon for a loan. As one Perth businessman said later: "People liked going to Laurie for money. They could go in at 9am and have their money by 9.15 with very little paperwork. The interest charges didn't matter in a bull market because they were making so much on the stockmarket with the money they got."

What nobody knew, however, was that between 1983 and 1987 Connell was siphoning off more than $130 million of depositors' funds from his firm, Rothwells, into his private interests. This had been hidden in Rothwells' glossy annual reports through a massive fabrication of the accounts. In the official report into the inevitable collapse of Rothwells, Malcolm McCusker QC, who is now the governor of Western Australia, found that the bank was effectively broke from 1985. "Depositors [of Rothwells] would surely have been alarmed had they known that they were, in reality, lending largely to Connell, on an unsecured basis, to finance personal acquisitions such as racehorse stables and art; and even more alarmed had they been informed that Rothwells, the company to which they believed they were lending, was effectively insolvent."

The freewheeling business culture of Perth in the 1980s was influential in shaping the young Andrew Forrest. By the middle of the decade, Bond and Connell were in full flight, the sharemarket was booming and Forrest was beginning to thrive. He left Benneys in 1985, aged twenty-three, for a job as a dealer with a new stockbroking house called Kirke Securities, which had a strong focus on the resources sector and boasted some big institutional clients on the east coast.

By this time, with his confidence soaring, Forrest was also able to bring his stammer under greater control as he worked the phones in search of new business. Just two years after being unable to speak in the presence of John Poynton, Forrest had earned the nickname "Silver Tongue" in Perth stockbroking circles for his powers of persuasion. "He was phenomenal at making phone calls," recalls Graeme Kirke, the principal of the firm. "He was always on two phones on the desk at a time – I'd never seen that before."

Forrest's time at Kirke Securities was brief but formative. Above all else, he learned the importance of establishing contacts with the biggest names in business. And the biggest name at the time in Australia was Alan Bond, who had become a client of Kirke Securities through a connection established by Forrest's colleague Brett Fogarty. (Fogarty himself would go on to make hundreds of millions of dollars in the mining boom of the 2000s through his engineering company GRD.) Forrest also wrote plenty of business for some of his wealthy connections in the pastoral industry. But the boy from the bush did take a while to adjust to life in the corporate world. Recalls Graeme Kirke: "We had to buy him a suit; we said to him, 'You can't keep wearing the chinos and the RM Williams.' He would come into work and there would be red dust stained from the seatbelt across his white shirt. Those sorts of things weren't an issue for him."

Although Forrest was unconventional, his peculiar brand of naïve enthusiasm won him admirers. One day the staff at Kirke Securities were working away in their first-floor office in central Perth when they were startled by a horrifying screech of brakes on the street below as a car slammed into a motorbike. Graeme Kirke remembers Forrest as the only employee who immediately ran downstairs to help. It was an impulsive act that, Kirke believes, underscores Forrest's strong sense of morality. "We all went to the window to have a look, but Andrew was the person who went straight down and got the guy off the bike. It was quite ugly and Andrew looked after him until the ambulance arrived. He does those things not because he wants anything from it; he does it because it's what you should do."

Fifteen years later, when he had become one of Australia's richest men, Forrest would perform a similarly impulsive act when he witnessed a vicious brawl between two men in the Perth CBD. According to media reports, the "incredibly brave" Forrest jumped in between the two men – both of whom were much bigger than him – and managed to break up the fight, but not before being walloped in the chest for his troubles. A television station showed mobile-phone footage of Forrest, dressed in a suit and tie, leading one of the men away from the brawl and calming him down, with the report labelling him a "good Samaritan" and a "humble billionaire".

In 1983, another significant event took place in Perth that would shake up the old regime: a youthful former television reporter called Brian Burke came to power as premier of Western Australia. Burke promised a fresh, entrepreneurial style of Labor government and immediately set about forming close alliances with the local businessmen who had long supported – and donated to – the Liberal Party.

Burke's intentions, at least initially, were pure: he wanted to boost the economy while rewarding Perth's business community – all for the greater good of the state's taxpayers. But his execution was deeply flawed. In what would become known as "WA Inc." Burke handed out contracts to his chosen tycoons in a series of dodgy deals that cost the state hundreds of millions of dollars. He would end up going to prison before re-emerging in the 2000s as a brilliant lobbyist for Perth's new breed of business leaders, including Andrew Forrest.

In the 1980s, Brian Burke saw something he especially liked in Alan Bond, a former sign-writer, and Laurie Connell, the son of a bus driver. Bond and Connell were certainly not the typical old-school-tie Perth businessmen who'd run the city for decades and they quickly sided with Burke. But the premier soon came to realise that Connell, who had been involved in a string of horseracing scandals and would end up going to prison, was hardly squeaky-clean. "I'd hate to stand between Laurie and a bag of money," he said.

Connell dazzled Perth with his conspicuous consumption. He owned 400 horses in Australia and England, hosted a grand reception for Princess Anne at his country estate and bulldozed six waterfront houses in Dalkeith, an exclusive riverfront suburb, to make way for a multimillion-dollar mansion on a new 10,000-square-metre "superblock" – a scheme that had to be abandoned when the good times ended in October 1987. It would be another two decades before Perth would again revel in the euphoria of a boom of such magnitude – and by that time Andrew Forrest would be richer than anyone who came before him. But in champagne-soaked Perth in the mid 1980s, few dared to ask whether Alan Bond and Laurie Connell might be crooks.

Forrest certainly wasn't about to question the business ethics of Bond or Connell. The tycoons would become two of his own treasured clients after he left Kirke Securities in 1986 to launch the Perth office of Sydney broking house Jacksons. Forrest was hired by Jacksons' managing director, Bob Pfafflin, who was hellbent on expanding the firm's operations in Australia and overseas at what was the peak of the 1980s stockmarket boom. Jacksons would become the first Australian stockbroking firm to be publicly listed, and it boasted branch offices in Melbourne, London, Paris, Munich and Hong Kong. Graeme Kirke says he wasn't surprised the precocious Forrest left his firm to accept Pfafflin's offer to run his own operation, even at the age of twenty-four. "I don't think Andrew could ever work without being the intellectual driving force," he says. Another of Forrest's colleagues from that era marvels at his audacity in taking on the establishment after he'd been in the industry for only three years. "That was something adults did – not 24-year-old blokes," he says.

The fit between Jacksons and Forrest was perfect. Jacksons had grown in the 1980s by backing many of the companies the establishment frowned upon, including Alan Bond's Bond Corporation. "Jacksons were go-getting blokes who wanted to take broking away from the old-school-tie environment," says former employee Peter "Cabbie" Richard. "They were never going to get business out of the conservative Collins Street in Melbourne, but as resources developed in WA, that created a whole new breed of people and they're the people that Forrest represented. Forrest was the start of the era where you had to be innovative, the resources industry was going up and you had to be in the middle of it."

At Jacksons, Forrest assembled a team of Perth whiz-kids who were fast-talking, hard-living and hungry for the next deal. The most colourful of the pack – which came to be known around town as the "Jacksons Five" – was the Ferrari-driving Dave Rigoll, who started out working as a plumber but got his break in the early 1980s when he was hired by Laurie Connell as a merchant banking junior at the ill-fated Rothwells.

Rigoll was often the last man standing, accompanied by a bevy of beautiful women, at the trendy Club Bay View in Claremont, where the cashed-up Jacksons boys would order "squadrons" of twenty-four drinks at a time. "Andrew idolised Dave," says a friend from that era. The super-slick salesman went on to make hundreds of millions of dollars through a Kazakhstani oil venture and at one point boasted houses in London, Dubai and the Swiss tax haven of Zug. And he never quite stayed out of trouble, attracting headlines in 2007 when he was implicated in a major share scandal at one of his London-based oil companies.

Forrest's next hiring at Jacksons would be Simon Lill, his old schoolmate and fellow prefect from Hale School. Lill, a talented athlete with a smart business brain, would become a key ingredient of Jacksons' success. Like many, he got carried away with the deal-making excitement of the era. After leaving Jacksons, Lill was investigated by the Australian Securities and Investments Commission and ultimately sent to prison for eighteen months on two counts of improperly using his position as a director. When he emerged from jail in 1997, Forrest gave him a plum job at his new venture, Anaconda Nickel.

Strangely, it seems there was never a fifth member of the Jacksons Five. For some time, Forrest even struggled to hire a fourth broker for his new enterprise. Jeff Braysich had been working at Patersons, one of Perth's most conservative and successful firms, having started in the industry in 1983, and was regarded as one of its best talents. When Forrest called him, Braysich rejected the job offer out of hand. But Braysich remembers Forrest repeatedly phoning him until he finally relented: "I wasn't sure whether to go, but he does wear you down. By the ninth phone call, I was saying, 'Okay, for fuck's sake, I'll come!' When I told Patersons I was joining Jacksons with Andrew Forrest, they were horrified." Braysich would later have his own brush with the law: he was sentenced to a twelve-month suspended jail term after being convicted of twenty-five counts of market rigging, involving shares in a Perth mining company in the 1990s, but after a long legal battle the conviction was eventually quashed by the High Court.

The Perth stockbroking establishment had never seen anything quite like Andrew Forrest and his crew at Jacksons. Forrest and Braysich were both just twenty-four; Rigoll and Lill were a year younger. But pretty soon, Perth's entrepreneurs were lining up to do business with these young turks, who traded feverishly in mining and energy stocks and would underwrite new floats worth millions of dollars. Every new resources explorer with dreams of making it big would come in to see Forrest. On several occasions, Jacksons hit the jackpot when a mining stock priced at 20 cents opened at $2 immediately upon listing. It was champagne all round on those days.

On the rowdy trading floor of the old Perth stock exchange on St Georges Terrace, the Jacksons boys were known as aggressive and cavalier. Regulation of stockbroking in the 1980s was weak; this was an era of few checks and balances. Insider trading was rife and, unlike today, there were no "Chinese walls" – the barriers within a firm that separate the research analysts, who provide advice to clients, from the traders, who do the buying and selling. John Poynton recalls many of the older brokers along St Georges Terrace frowning upon Jacksons. Jeff Braysich agrees: "We weren't exactly loved around Perth. If we were invited to functions, they weren't clamouring to associate themselves with us. It was because of the nature of the deals we were doing and the publicity we were generating."

Besides Bond and Connell, Jacksons' key clients in Perth included Robert Holmes à Court (the legendary corporate raider who tried to take over BHP), Yosse Goldberg (a key player in the WA Inc. scandal whose company, Western Continental, sank in 1987 before he fled to Spain) and Mark Povey (the youthful Rolls-Royce-driving tycoon who was bankrupted when his petrol empire fell over). Forrest also established strong links with colourful Melbourne mining tycoon Joseph Gutnick and Sydney insurance king Larry Adler and his son Rodney (another of his good friends who would later be sent to jail). It was the most spectacular bull market anyone in the industry had seen, and the biggest names in Australian business lined up to trade the mining stocks being pushed by Jacksons. "They did their big deals through the big brokers and they'd give the smaller stuff to the small brokers like us who were hungry," Braysich recalls. "We were used by all these entrepreneurs – they inflated our egos to get us to do what they wanted to."

Even at a young age, Forrest was keen to show his colleagues who was boss. But he must have been taken aback when the Perth stock exchange – for reasons that remain unclear – signalled it would not accept the young firebrand as a member. Forrest reluctantly agreed that Braysich, who had an MBA and was more acceptable to stockbroking traditionalists, could be the firm's representative so it could open its doors. Forrest's version of the story, as told to friends, is that he was never rejected and Braysich was chosen to be the firm's representative simply because he was already a member of the exchange.

Braysich says this created friction between the pair over seniority. The issue came to a head one day when a journalist from the *West Australian* arrived to interview Braysich about the new firm. "Andrew comes straight in and says, 'Hello, I'm Andrew Forrest' – he was wearing moleskins, the navy jacket with the gold buttons and the blue chammy shirt," Braysich recalls. "He got really upset because I was stealing the limelight."

Reflecting on the 1980s, Braysich says he and his colleagues at Jacksons were flamboyant because they were so young and had only ever known an industry in boom times. "We didn't know any better – we thought this was what it was like; we didn't have ten years of tough times. The older guys in the industry were more cautious and sober," he says. Braysich says he admires some of Forrest's achievements since Jacksons but acknowledges he was difficult to work with. "You were either his best mate or worst enemy, and at various times I was both," he says. "He's got a hide on him. You could shoot him in the arse and he wouldn't know. He is as brash and as cocky as anyone you'll ever meet."

It was around 1986 that Forrest struck up an important friendship with Woody Pearce, an old Kimberley pastoralist who had struck it rich by investing in gold stocks. Pearce had three daughters, one of whom worked at Jacksons as a personal assistant. Forrest had become romantically involved with one of Pearce's other daughters, Merrilee. It was a tangled web at the time, because his mother, Judy, by then divorced from Don, later started a relationship with Woody Pearce. Woody became Andrew's stepfather when he married Judy.

When Woody died in the 1990s, Forrest fell out badly with the Pearce daughters over the will. The daughters believed their mother, Georgina, had not been looked after, and they blamed Forrest, who emerged with Woody's Mercedes and other items (Forrest has told friends he bought the car from the estate). To Andrew, Woody was an inspiration. Those who knew Forrest in the late 1980s say Pearce taught him how to read exploration reports and had a strong influence on the way he viewed the mining industry. Others say Pearce shaped Forrest's whole approach to life, infusing him with the belief he could achieve anything if he dreamed of what most people considered impossible. Forrest also traded stocks for Pearce. "He was a mentor and he perhaps helped open Andrew's eyes to the possibilities in the sharemarket," says one old friend. "Woody was a thinker."

If there was one thing at which Jacksons really excelled, it was partying. Forrest was renowned for his ability to work at least twelve hours a day, drink late into the night and still be back at his desk the next morning. "He was happy to go out until four in the morning and get to work at seven and still function; his mind is always active," says an old colleague.

Jacksons' trips to the WA Goldfields, often with fund managers and journalists in tow, became legendary among the stockbroking crowd. The firm's Melbourne-based head of corporate, Ross Dobinson, once arm-wrestled a huge red-bearded local miner in the front bar of the Leonora Hotel; somehow, he won. Another time, the well-lubricated Jacksons boys drove a fire engine down the main street of Mount Magnet (a donation to the local brigade the following year was intended as an apology for that indiscretion). And on the bus trips between Kalgoorlie and Perth, the brokers would challenge themselves to drink more litres of Emu Export beer than the vehicle used in fuel. The most sober one on the bus would be responsible for keeping a tally.

The money really started pouring in at Jacksons in 1987, allowing Forrest to buy a gleaming new Jaguar. Perhaps he was simply trying to keep up with his colleagues, who also enjoyed spending their newfound wealth on fast cars. One day in the Jacksons office, Dave Rigoll asked Simon Lill to drive him to a car yard so he could pick up the new Ferrari he had ordered. But by the time the pair arrived, Rigoll had managed to convince Lill to buy one for himself as well. Lill immediately wrote out a cheque and the two brokers headed back to the CBD to parade their new toys.

This sort of madness could not last, of course. The bull market that had been fuelled by cheap cash and greed ended abruptly in Australia on 20 October 1987, when the sharemarket lost one-quarter of its value in a single day of panicked selling. Most of the stocks Forrest had been spruiking to clients were now pretty much worthless. He has since told colleagues that he personally lost $3.5 million in the crash. It wasn't money he had in the bank, but the paper profits he'd built up as the market surged.

Forrest had to endure plenty of angry clients calling to ask what had happened to their money. But one of his more highly valued clients, Rodney Adler, was particularly unhappy to discover that he had lost several hundred thousand dollars on a stock he had never wanted in the first place. Adler was then the investment manager for his family insurance company, FAI, and he had been impressed with Forrest's zeal for stockbroking. But the value of Adler's investments had plummeted in the crash.

Despite Forrest's financial worries at the time, he was never downbeat. A day after the crash, the *Australian Financial Review* asked several Perth stockbrokers whether they thought the market could recover anytime soon from the biggest collapse in its history. All of them – apart from one – were bearish about the outlook. The sole optimist was Forrest, who blamed the fall on the herd mentality of Australian investors and predicted the punters would come rushing back into the market to buy undervalued stocks. "There are some excellent situations offering value now and we are recommending buys," said Forrest. Unfortunately for Forrest and the cheap stocks he was flogging, the Australian stockmarket continued to fall for several months and would not return to its pre-crash peak for another six and a half years.

With the Perth market virtually blown away overnight, there wasn't much more Forrest could achieve in the wild west. It was time to move to Sydney. If truth be told, Bob Pfafflin and his fellow directors at Jacksons had already grown slightly uncomfortable with the "cowboy" reputation of the Perth office, so transferring Forrest to Sydney was a way of keeping a closer eye on him. But the directors had another motive: Forrest had made plenty of money for Jacksons in the boom and they hoped his talents might help lift the firm out of its post-crash despair.

# 5.

# SIN CITY

*Any normal, sane person wouldn't have done it.*
*But with Andrew, it's no risk, no reward.*
—ALBERT WONG, Forrest's former business partner

Within minutes of first meeting his Jacksons colleagues in Sydney, Forrest had left a lasting impression. It was 1986 and the big-talking hotshot from Perth was in Sydney to work from the firm's head office for a few days. The seventy brokers had just finished their daily meeting at 8.30am when a typically effusive Forrest picked up the phone and began speaking loudly in a bid to impress his new workmates. Calling himself "Kookaburra", Forrest had phoned Laurie "Last Resort" Connell in the hope of drumming up some business. But the call lasted only ten seconds. The broker sitting next to him at the time, Peter "Cabbie" Richard, turned to Forrest to ask whom he had just been speaking to.

"It was Laurie Connell," Twiggy said. "What did he say?" asked the gregarious Richard. A somewhat sheepish Forrest replied: "He said, 'It might be 8.30 in Sydney but it's bloody 5.30 in the morning in Perth, don't ever bloody ring me at 5.30 in the morning again or I'll come over there and cut your balls off!'" Nevertheless, Connell called back a few hours later and placed a trade with Forrest.

When Forrest moved to Sydney permanently in early 1988, Richard was tasked with ensuring the young gun's energy was being channelled effectively. Forrest's style, as usual, was crash through or crash. If a group of brokers in the office happened to be talking about BHP, Forrest would get on the phone straight away and demand to speak to BHP's chief executive. Most of the time he was unsuccessful but nobody could doubt his verve. "The normal procedures you'd follow to do things, Forrest never did," Richard says. "He's the sort of bloke who wanted to pick up whatever he had in his hand at the time and run with it at 100 miles an hour in whatever direction he could. He wasn't the best broker I'd ever seen but he was out there, he was in your face. He was a better entrepreneur than he was a broker, ultimately."

Another colleague, Ross Dobinson, remembers Forrest as a skilled raconteur whose workmates were enthralled by his relentless enthusiasm for stockbroking. Forrest would regularly hold court over long lunches at the Imperial Peking Harbourside restaurant, a short stroll from Jacksons' Spring Street offices in the heart of the Sydney CBD. "A lot of people criticise him for his enthusiasm," Dobinson says. "But we were all growing up back then and he was just more successful at being enthusiastic than anyone else."

Forrest had been buying up shares in Jacksons since the firm listed on the stock exchange in early 1987, and by the end of that year he owned 10 per cent of the company. That made him the biggest shareholder of a firm in which he was still a relatively junior employee – an extraordinary situation by any measure. But in May 1988, with the market still in the doldrums and Jacksons losing money hand over fist, Forrest quit the firm and sold all his stock. He even managed to turn a tidy profit on the investment, selling shares he picked up for 75 cents at $1 each. It was exceptional timing: Jacksons collapsed a year later, owing the National Australia Bank $3 million and carrying bad debts of $11 million.

The thrillseeking broker craved a fresh adventure. And what better way to satisfy the urge than by going into business with Alan Bond? The tycoon was one of the few still standing in the wake of the stockmarket meltdown, although in reality his empire was stretched to breaking point and would begin to fall apart within two years. Forrest formed a broking and merchant banking venture, Intersuisse, in a 50–50 partnership with an investment company, Markland House, which was owned by Alan Bond, Perth businessman Brian Coppin and Sydney banker Julian Hill. But as would become apparent to investigators only when they were able to comb through Bond's financial dealings, the entrepreneur was using Markland House as a conduit to strip another company he controlled, Bell Resources, of more than $1 billion to prop up his struggling Bond Corporation. Bond would go to jail in 1997 for the enormous fraud.

An association with "Bondy" in 1988 still carried kudos for someone like Forrest, however. Ever the optimist, he outlined plans for Intersuisse to poach key managers from rival broking houses and to appoint institutional and corporate dealing teams in Perth, Melbourne and Sydney. They were bold words at the time because the market was dead in the water and most other broking firms and banks were laying off staff.

Despite the huge paper losses Forrest had suffered in the stockmarket crash, life in Australia's biggest city was sweet for the knockabout kid from Minderoo. Forrest had branched out on his own at Intersuisse, he owned a bachelor pad in the harbourside suburb of Darling Point and he had enough spare cash to reward himself with a Harley Davidson to navigate the city streets. Twiggy attracted women easily and was rarely without a girlfriend. His sister, Janie, had also moved to Sydney and was running a Kings Cross hotel that became his regular haunt for endless rounds of cocktails.

A close friend from that period, David Hannon, estimates Forrest made at least forty good friends in his first year in Sydney, most of whom were not stockbrokers. One of these mates was

"Rocket" Rodney Adler, the brash businessman for whom Forrest had traded stocks while at Jacksons. Adler recalls that he enjoyed going out on the town with Forrest, but he could never match the Western Australian's drinking prowess. "I did not want to, nor could I, drink that volume," he says.

In 1988, Forrest met Albert Wong, who was working for another firm in the same Bligh Street building as Intersuisse. Wong is an affable investment banker with slicked-back hair and thirty years' experience in wheeling and dealing in Sydney, often in partnership with former NSW premier Neville Wran. Soon after the pair met, Forrest was lobbying Wong, then in his mid twenties, to join him at Intersuisse. Wong could see some potential in working with Forrest, but he declined. As many others have learned, however, Forrest doesn't take no for an answer. "Andrew with his silver tongue just continued to persevere, so I joined," Wong says. Wong later took a 20 per cent stake in Intersuisse, leaving Forrest and the Alan Bond-backed Markland House with 40 per cent each.

As a banker, Wong had hitched his fortunes to one big client, Japanese hotel group EIE International, which was emerging in the late 1980s as one of the largest foreign investors in Australia. Forrest was keen for exposure to what he believed to be a looming flood of Asian investment into the country. With Wong on board, one of Intersuisse's first major deals involved EIE buying Macquarie Bank's stake in a floating luxury hotel that was perched off the Great Barrier Reef. Forrest and Wong had to negotiate with government agencies and shipping companies to tow the huge structure to Vietnam on behalf of EIE. Their relationship with EIE was lucrative, but by the early 1990s the Japanese company had collapsed with $11 billion in debts, and its boss was prosecuted for fraud in Tokyo. EIE wasn't to be their only client that ultimately went broke or ran into legal troubles. Others included corporate raider Bruce Judge and dodgy property developers Mike Gore and Malcolm Edwards.

Forrest and Wong did deals around the clock and kept fit by jogging around Sydney's Nielsen Park after work before heading out for drinks. In defiance of the general economic malaise, Intersuisse grew rapidly to have more than fifty staff in six offices, including in London, Auckland and Taiwan, where Forrest first caught a glimpse of the Chinese entrepreneurial spirit. And when the young deal-makers travelled overseas, they always flew in style. "Andrew taught me how to fly first-class," Wong recalls. "I remember the cheapest first-class ticket to London was on Air India, which nobody used to fly. Andrew and I got into the first-class cabin and there was only one other person. They minded their own business and we got plastered all the way to London."

One day Forrest and Wong "borrowed" EIE's Boeing 727 corporate jet and took it across the country to Minderoo for the weekend, flying low for a few loops around Uluru on the way to the Pilbara. At the Forrest family station, Wong witnessed his mate's constant pursuit of the quick thrill when he jumped into the muddy Ashburton River from the wobbly branch of a tree that was at least ten metres off the ground. "Any normal, sane person wouldn't have done it," he says. "But with Andrew, it's no risk, no reward."

One of the personal assistants employed at Intersuisse's office was a quiet, refined woman by the name of Nicola Maurice, who hailed from a Brethren farming family in rural New South Wales and had studied economics at university in Canberra before moving to Sydney. To observers, she was the antithesis of Andrew Forrest, the party-boy stockbroker who gave the impression of worshipping Mammon more than Jesus. Albert Wong admits he was surprised when Forrest declared to him in 1989 that he had fallen for the blonde girl he had nicknamed "Sweet Pea". Wong recalls: "One night he just said to me, 'I'm going to take Nicola out,' and I said, 'You're kidding!' Romance just blossomed from there."

Andrew and Nicola became engaged but Forrest soon got cold feet and broke it off, mainly because he wanted to remain a bachelor for longer. "He had a real weakness for the ladies," says an old friend. A heartbroken Nicola quit her job and fled to Europe, where she took a job as assistant to the finance director at the United Nations in Geneva. Andrew's stepmother Marie, Don's second wife, explained this episode to journalist Cameron Stewart in 2008: "He got a bit tired of talk of bridal dresses and babies and he said, 'I don't think I'm ready to be married, still a few more girls in the field.' I said to Andrew at the time, 'You have just got rid of the best part of yourself. She is such a lovely girl.' About nine months later he said, 'Well, I guess if I want to marry little Nic I'd better go and chase her.' He did that, but from then on it was on her terms."

Forrest flew to Europe, ostensibly to do some work for an oil industry client, which paid his airfare. But his real mission was to find Nicola and ask her to take him back. She agreed; however, she was determined to instil more Christian principles in her wayward fiancé and demanded the couple do a counselling course before they walked down the aisle. Her deep faith in God was a big influence on Forrest, and friends noticed he started to become more pious. Nicola began taming Andrew in a way no one else had been able to. As Don Forrest said in his matter-of-fact way: "Nic has been very good for him; she has taught him other values."

As Forrest was sorting out his romantic life, his business relationship with Albert Wong began to unravel. In late 1989, Forrest took a holiday to Africa that was originally meant to last for three months but was extended to allow him to be in South Africa for the release from prison of Nelson Mandela in February 1990. "When he was away all the skeletons started to come out of the closet," Wong recalls. One of the firm's backers, Julian Hill, ran into financial strife and Intersuisse found itself in breach of the liquidity ratios it needed to remain a member of the stock exchange. Wong urgently borrowed money to buy out the stake in the firm owned by Hill, Coppin and Bond. That gave him 60 per cent of the company and therefore the upper hand over Forrest. Intersuisse, which had expanded too quickly in a weak economic climate, was teetering on the brink of failure. In the days before mobile phones and email, Forrest was unaware of much of the drama that was unfolding back in Sydney.

When Forrest finally returned from Africa, Wong presented him with an ultimatum: either you buy me out of this mess or I will buy you out. According to Wong, Forrest didn't want to be pushed out of Intersuisse because he had lent the company $300,000. Things turned nasty when Forrest took Wong to the NSW Supreme Court. "Andrew's natural instinct is always to fight," says Wong. But the case never made it before a judge, as Forrest and Wong reconciled at the last minute, reaching a settlement on the steps of the courthouse. Forrest sold Wong his stake in 1990 for a much smaller sum than he had hoped for and went off in pursuit of his next wild plan: breeding alpacas.

Alpacas are lethargic beasts that look like a cross between a llama and a small camel. Until the 1980s, they were bred for their silky-soft fibres only in their native South America. Forrest's idea was to bring the alpacas from the highlands of Chile all the way to Australia – where they were worth $20,000 each – and to ultimately sell them to the North Americans and Europeans, who were prevented by strict quarantine laws from importing the livestock. But first he needed to raise millions to buy hundreds of the animals. Forrest was quick to pounce on a recent change in the law in New South Wales that allowed for limited partnerships in agricultural schemes. This meant that investors – usually wealthy lawyers, doctors and other professionals – could put $10,000 each into the scheme but would only ever be liable for a fraction of their original outlay. Their entire upfront investment could also be written off as a tax deduction.

Forrest formed a company called International Alpaca Management and sought to raise $5 million, promising investors that the alpacas would deliver a final payout of up to $49,000 each. While that sounded like an outstanding return, only about $3 million of the $5 million arrived. Rodney Adler's FAI ended up putting in money to help meet the shortfall and the scheme was off the ground. Forrest became obsessed with the scheme, which took him back to his agricultural roots in the Pilbara. With his mother, Judy, and sister, Janie, he established a breeding stud near Mittagong in the NSW southern highlands. Forrest might have been living in Sydney's fast lane, but now he had the rural retreat – and a Range Rover – to lend credence to the country gentleman persona he liked to cultivate.

The alpacas were shipped from Chile to the Cocos Islands, the remote Australian territory in the Indian Ocean where the government had a quarantine station. But that's when Forrest's new get-rich scheme ran into trouble. While about 100 of the animals were waiting to be cleared to enter the Australian mainland in late 1992, a dispute over ownership erupted between Forrest and an intermediary in the deal, New Zealand businessman Ben Ensor. Forrest responded by suing Ensor, and what followed was a lawyer's feast. A flock of highly paid QCs lined up for hearings in the Federal Court that stretched over several months in 1993 and 1994.

When Justice Margaret Beazley finally delivered her ruling in April 1995, she became the first of several judges to shine an uncomfortable light on Forrest's business methods. Beazley found that although Forrest had attempted to answer questions during the trial honestly, he "demonstrated that he would conduct his business in a way to achieve his commercial ends, even if that involved threats and falsehoods". The judge was referring to evidence Forrest had given in which he admitted that claims he made in a letter to Melbourne barrister Chris Wren, an investor in the scheme, were not what he believed at the time but were justified because they were for "negotiation purposes".

Beazley also found that Forrest presented in the witness box as "self-assured and at times amused by his own evidence". It was not a good look for Forrest, who by the time the judgment came out was about to become the chief executive of Anaconda Nickel, a listed public company. But the judge's comments did not shock many of those who had encountered the slick salesman in business dealings over the years. Forrest strongly rejected Beazley's comments in an interview in 2001: "Anyone who knows me will tell you I'm a passionate, enthusiastic individual, but I'm not a person who gets through life through threats and falsehoods." Forrest lost the first round of the alpaca case, but he appealed the ruling and won. Ben Ensor, who by that stage was representing himself, then asked the High Court for leave to appeal, but this was rejected in 1996. Victory was Forrest's at last.

The alpaca scheme may have had its difficulties – and been superseded in Forrest's hyperactive mind by bigger enterprises – but the breeding stud at Mittagong survives today and is still run by Janie. Adler says FAI managed to recover its money and even made a small profit but the business venture was "quite marginal".

After leaving Intersuisse in 1990, Forrest began looking around for another job in the financial world. He picked up some freelance work negotiating on behalf of local entrepreneurs who were in trouble with the banks during the recession of the early 1990s. And he began spruiking an audacious proposal to create a $100-million trust to invest in goldmining ventures. It was a typical Forrest plan to take on the world, although the weak markets at the time prevented it from ever happening.

During this period, he met Warwick Grigor, a thoughtful 32-year-old who had earned a law degree but become a leading mining analyst at investment bank County NatWest Securities in Sydney. Grigor had, by chance, been assigned to help Forrest with the gold trust he was trying to get off the ground. When he was made redundant in April 1991, he phoned Forrest, who asked Grigor to meet him later that day at the Royal Hotel in Paddington. Over a few schooners, Forrest pitched the idea of setting up a company to advise and finance small miners. Grigor would be the analyst, Forrest would be the salesman and they'd both become rich beyond their dreams. "Everyone advised me not to do it – they didn't trust Andrew," Grigor says. "But what I liked about Andrew was his positive style. He was obviously smart. I didn't know how successful it would be, but I said, 'Yeah, why not?'"

For the conservative Grigor, it was to be the wildest ride of his life. The pair launched their new investment bank, Far East Capital, in the basement of Forrest's newly acquired house in Suffolk Street, Paddington. Money was tight. They bought desks and office equipment for the business from auctions they would attend in The Rocks. At the end of the first year, they had paid themselves only about $20,000 each in salaries. Grigor says Forrest was heavily distracted for the most of the time at Far East Capital by his alpaca scheme. "In the first three years I brought in most of the income," he says. "It was my hard work that made the money because he was off playing with his alpacas. It wasn't until Anaconda came along that he started to deliver."

Forrest did, however, cultivate some wealthy young clients at Far East, including British financier Nick Barham, who at the time was the son-in-law of the nation's richest man, Kerry Packer. Forrest later became friendly with Packer's son, James, attending his $10-million wedding to Jodhi Meares in 1999 and his second wedding to Erica Baxter on the Côte d'Azur eight years later, even drinking shooters around the hotel pool with the groom on the day before the service. Another scion of the Sydney establishment, Rodney Adler, who at age twenty-nine had stepped up to succeed his late father as the head of FAI, was also a client of Forrest's at Far East. For years afterwards, Forrest would be able to boast of his close links to both the Packers and the Adlers, two of Sydney's best connected families.

Forrest also helped position the bank to focus on the rise of China, which was taking tentative steps in the early 1990s towards opening up its long-dormant economy. Albert Wong remembers speaking to Forrest after he had set up the new venture. "He said to me, 'Albert, you know the future is Asia. Come to my office, I've even got an Asian girl as a receptionist. And that's why I'm called Far East Capital.'"

As the firm started to do more deals, the cash came rolling in. Just as they had planned at the Royal Hotel a few years earlier, Forrest and Grigor became rich. Andrew and Nicola, meanwhile, got married in October 1991 in a relatively low-key affair at the NSW southern highlands property owned by Nicola's parents, Tony and Brooke Maurice. Forrest's best men were his brother, David, and his old schoolmate John Morrison.

The Maurices, however, harboured a secret of which few of the wedding guests were aware: they were key figures in the Australian League of Rights, a far-right Christian group regularly accused of anti-Semitism. In 1991, as Andrew and Nicola exchanged vows, the Human Rights and Equal Opportunity Commission issued a report describing the League as "undoubtedly the most influential and effective, as well as the best organised and most substantially financed, racist organisation in Australia". Tony and Brooke Maurice had been hosting meetings for the League since the early 1970s. A front group created to contest the 1983 federal election, the Christian Alternative Movement, was formed at the farm they owned at the time in western NSW. In the 1980s, Nicola's elder sister, Katrina, married David Thompson, who went on to become the League's national director. Thompson told League members upon his appointment in 1992: "In order for the Christian faith to command its place in the sweep of history, its practical application in every field of human endeavour is essential." He also courted controversy by organising Australian speaking tours by the British academic David Irving, the infamous Holocaust denier.

Andrew Forrest may also have fallen under the League's spell. In 1994, his colleague Warwick Grigor was taken aback when Forrest invited him to attend a League of Rights meeting in Sydney. Grigor recalls he told Forrest it "probably wasn't a good idea" to attend such a meeting because of the League's poor public image. To this day, he can't be certain of whether Forrest had any involvement with the group. Forrest's apparent interest in the League's activities does not suggest he held any racist views. As the HREOC noted in its 1991 report, the League's extremism was concealed by its espousal of family values, patriotism and nationalism – all qualities Forrest has long embraced. "The League represents the respectable face of racism. Its advocacy of traditional values may have won it mainstream support from people who are unaware of its racist and extremist ideals," the report said. Perhaps Andrew and Nicola were attracted to the League's evangelistic Christian courses, which were run at the time by Jeremy Lee, a key figure in the Christian Alternative Movement launched at Nicola's childhood home.

Forrest got on well with his brother-in-law David Thompson. When Thompson stood down as head of the Australian League of Rights in the late 1990s, Forrest even hired him as the external relations manager at Anaconda Nickel, a senior role that involved lobbying the company's powerful shareholders in a bid to keep Forrest on the board. Thompson, who had learned plenty about politics and campaigning during his long association with the League, remained with Anaconda until Forrest was eventually forced out in 2001. Today, he insists he doesn't recall Forrest ever being personally involved in the League, but he says they have remained in touch and share a commitment to Christianity.

By the time they were married, Andrew and Nicola were mixing increasingly with Sydney's wealthy eastern suburbs set. Their first child, Grace, was born in mid 1993, just six weeks after they bought federal Liberal Party leader John Hewson's mansion in the upmarket suburb of Bellevue Hill for $1.02 million. It was an impressive amount for a relatively unknown 31-year-old banker to pay, considering the median house price in Sydney that year was just $188,000. Today, the Fairfax Road property is worth more than $4 million. Built in 1900, the sprawling two-storey house has five bedrooms, a dining room with seating for twelve people, a morning room, a den, a formal sitting room and a piano room. Outside are parking bays for several cars, a swimming pool and a self-contained apartment. To top it off, the house is just 250 metres from Sydney Harbour and a relaxed stroll to the plush boutiques of Double Bay.

Ever the promoter, Forrest was quoted in the *Australian Financial Review* in 1993 as saying he believed his new digs would prove to be a sound investment since houses around the $1 million mark represented the "softest" part of the Sydney market. As it turned out, Forrest never made the big gains on the property he had predicted, selling it three years later for $1.16 million – or around 14 per cent more than what he paid.

Looking back, Grigor reckons the accumulated stress of working with Forrest for five years took its toll on him and contributed to some major health problems. Many of the tensions between the pair at Far East – and plenty of their successes also – arose from the fact that Grigor and Forrest were polar opposites.

While Forrest was partying with his wealthy Sydney mates, Grigor was happily ensconced at his family home in suburban Sutherland, twenty-five kilometres south of the CBD. When Forrest was out zealously spruiking the latest mining stocks, Grigor was carefully poring over data in balance sheets. Forrest's main mission in life at that point was to make pots of money, but Grigor, who speaks Gaelic and plays the bagpipes, says that has never been his main motivation.

Grigor says he spent a lot of time dealing with the Australian Securities Commission and the Australian Stock Exchange about Forrest. "He had a fairly cavalier approach," Grigor recalls. "With my legal training I'd have to come in later and sweep up after him a fair bit." In 1996, when Forrest became chief executive of Anaconda Nickel, he said publicly he'd be stepping back from his role at Far East Capital to concentrate on the new job.

Within months of that 1996 announcement, however, Forrest had decided to leave Far East altogether and return to Perth to run Anaconda, a company he had been advising for three years. It was a big punt but he was convinced Anaconda was a pathway to great wealth. He offered Grigor the role of chief financial officer of Anaconda, which Grigor turned down. "At that stage Andrew was a little bit out of control and I didn't want to work for him," Grigor says. "I didn't want to be an employee because he'd be too unreasonable – it was challenging enough having him as a partner. The liberties he would take would conflict with my conservative side."

The split with Grigor was amicable at first but soon turned nasty after a dispute erupted between the pair over the ownership of a pile of Anaconda shares worth millions of dollars. By now, these sorts of clashes were becoming common for Forrest. Just as he had fallen out with Albert Wong at Intersuisse in 1990 and then become embroiled in a bitter legal fight with his associates over the alpacas, so too his once-close relationship with Grigor fell apart in late 1996.

When Forrest left Far East Capital, the assets of the firm were split between the two partners. These included 9 million options in Anaconda Nickel that could be converted into shares if the stock price hit 30 cents. Grigor says he exercised his 4.5 million options and sold the shares in April 1996, when the stock was trading at about $1.30, pocketing a cool $6 million. But the shares soared to $4.50 within six months, meaning Grigor had potentially missed out on a much bigger payday. Grigor claims he was convinced to sell his shares by Forrest and his fellow Anaconda director Michael O'Keeffe, the local boss of Swiss commodities trader Glencore, which had invested in the nickel company.

Grigor claims Forrest had long been angry that the two men had the same number of shares in Anaconda even though Forrest had done most of the work in advising the aspiring nickel producer. To this day, Grigor has a bad taste in his mouth over his decision to sell the shares at $1.30 just before they soared.

A separate disagreement between the two men over Anaconda shares ended up in the NSW Supreme Court in early 1997. Grigor will say only that the pair resolved the dispute over a bottle of scotch, and he is no longer bitter towards his old partner. They even catch up occasionally for a chat these days. "Andrew had a favourite term, 'Let's bury the hatchet,'" says Grigor, who credits Forrest with teaching him plenty about business in the 1990s. "He doesn't like to bear grudges."

# 6.

# DREAMS IN THE DESERT

*I've met a lot of managing directors, but I don't think
I've met a more self-evidently enthusiastic and dedicated
managing director than I've met in Andrew Forrest.*
—John Howard

Andrew Forrest got involved with a two-bit company called Anaconda Nickel because nobody else dared to touch it. In 1993, two Perth mining identities – prospector Peter Salter and metallurgist Geoff Motteram – were desperate to find someone who would underwrite a public float of the company, which had some old tenements in the burning desert north of Kalgoorlie. After failing to convince almost every financier in Perth to back their plan, Salter and Motteram flew to Sydney to try their luck with Forrest at Far East Capital.

Anaconda's problem at the time wasn't Motteram, who was respected in the industry after stints with big mining companies, including BHP and Western Mining Corporation. Rather, it was the presence of the burly Salter – or "Salty Pete", as he was known around Perth. Salter had a colourful history in the city's business and horseracing circles, having been declared bankrupt in the 1980s. He would later be banned by the corporate regulator ASIC from serving as a company director for three and a half years for his role in a failed bottled-water venture. Salter returned to business only to be sacked as chief executive of a

Perth-based gold explorer over a sexual harassment case a few years later.

Always keen on a gamble, Forrest was willing to overlook Salter's tarnished reputation and embraced his plan to develop mining leases near the abandoned town of Murrin Murrin, in Western Australia's northern Goldfields. The relationship between Forrest and Salter would break down two years later in a haze of drunken punches, but for a brief time the pair shared a common vision of creating a wealth-generating oasis on the edge of the Great Victoria Desert.

Reclining in his office chair, Forrest told Salter and Motteram that Far East Capital would underwrite the $6-million float of Anaconda. The money raised from investors would be spent on drilling tenements that had been pegged during the 1960s nickel frenzy by a company called Australian Selection, which eventually discarded them due to the frighteningly high costs and technical complexities involved in extracting what are known as nickel laterites from the ore.

Up until that point, all nickel in Western Australia had been produced from sulphide ores, which are relatively high-grade and quite simple to process before being sold. The lower-grade laterite ores are weathered rocks found near the surface, which are easier to mine but far more difficult to treat. As a metallurgist, Geoff Motteram was aware of a little-understood extraction method for nickel laterites called high-pressure acid leaching (HPAL) which had been pioneered in the 1950s at Moa Bay in Cuba. American mining giant Freeport built the Cuban plant but the project disappeared from public view when Fidel Castro seized control of the country in 1959, nationalising the nickel industry and naming one of the Moa Bay plants "Ernesto Che Guevara" after the Marxist revolutionary hero. For the next thirty years, the consensus in the West was that the Cuban plant, which relied on Soviet replacement parts, was a dud. It wasn't until 1994 that Canadian company Sherritt International, a specialist in HPAL, was able to invest in Moa Bay and begin to modernise the plant.

In 1994, Andrew Forrest was also launching his sales pitch for his $1-billion plan to build the world's first HPAL plant outside Cuba. Forrest's message might have baffled a few investors, but it went something like this: Anaconda will build a nickel mine; the ore from the mine will be crushed and fed into a colossal processing plant; it will then be mixed with sulphuric acid at temperatures of up to 250 degrees Celsius at high pressure inside four huge boilers known as autoclaves; the nickel and cobalt will then be leached out; and the end result will be small, pillow-shaped briquettes that will pop out the other end of the plant. The high-grade nickel and cobalt would then be shipped around the world to make stainless steel, jet engines, batteries and other necessities of modern life.

Besides his promotional zeal, Forrest brought to Anaconda a strong financial awareness and a vision for the bigger picture. He understood that the world was running out of traditional sulphide ores, but also that the future industrialisation of Asia meant nickel would always be in demand. And he came to believe that the desert sands around Murrin Murrin contained a bounty of nickel laterites for anyone prepared to explore there. Forrest also realised that one of the most important assets for an energy-guzzling nickel plant was access to cheap gas. And he knew that Richard Court's newly elected Liberal government in Western Australia was promising to build a gas pipeline from the north-west of the state to supply the mining operations of the Goldfields. The pipeline arrived in the Goldfields in 1996 just as Anaconda was preparing to start construction.

Forrest's earliest backer at Anaconda was Rodney Adler, who by the early 1990s was a major figure in Australian business. Forrest met Adler one day to tell him about the nickel tenements in the Goldfields and the gas pipeline. Adler recalls: "He came to me and said, 'Rod, I have the greatest idea I've ever come up with – I want to be a nickel producer. The pipeline will come through, we'll raise some money and we'll start an operation. I'm telling you, it's going to be one of the best things we ever do.'" Adler was sold on the idea and put some money in on behalf of FAI. He would eventually invest $4 million and end up with a profit of $65 million. The FAI stake would also provide Forrest with a crucial partner in later years as he battled other shareholders for control of Anaconda.

Anaconda had plenty of sceptics in the early days. Many believed that the ore body at Murrin Murrin wasn't good enough and the extraction technology too complex and unproved. Others said a glib ex-stockbroker with no real mining industry experience would never be able to raise hundreds of millions of dollars from serious investors. The nickel industry in the 1990s was dominated in Western Australia by Western Mining Corporation, whose chief executive, Hugh Morgan, was the embodiment of the conservative business establishment. WMC, which had banked its future on traditional sulphide deposits, let it be known that it doubted whether Forrest could make Murrin Murrin work. As he would do again several years later in the iron ore industry, Forrest cast himself in the role of underdog against the might of the established nickel players.

The history of nickel laterite projects, both in Australia and overseas, has long been a chequered one. Although the Greenvale operation in Queensland had been a success since the 1970s, two other projects launched in Western Australia in the 1990s, Cawse and Bulong, were big disappointments. In 2009, even the world's biggest mining company, BHP Billiton, couldn't make money out of nickel laterites, losing billions of dollars after walking away from its Ravensthorpe plant in the south of Western Australia.

But in the mid 1990s, Forrest had a story to tell. He declared Murrin Murrin would become the world's fifth-biggest nickel operation, producing 45,000 tonnes of the silvery-white metal a year and feeding the world's rapidly growing stainless steel markets for decades to come. What's more, Murrin Murrin would deliver shareholders juicy profit margins because it would be producing nickel for just $US1 a pound, well below the then market price of $US4 a pound. The dream also appealed to Forrest's strong sense of a family destiny. John Forrest had ridden over the Murrin Murrin site in 1869 on his first inland expedition, even naming a nearby prominent hill Mount Margaret after his future wife.

Forrest worked obsessively at drumming up interest in the Anaconda float, telling potential investors that the 20-cent shares were certain to soar to more than $1 in virtually no time (in fact, the shares would take two years to get to that point). Sure enough, Anaconda raised the $6 million and was listed on the Australian Stock Exchange in March 1994. Forrest installed himself as chief executive the following year and stacked the board with his allies, including his sister Janie's husband, Aldous Hicks, a one-time candidate for Liberal Party preselection against John Hewson in the federal seat of Wentworth. Forrest asked his accountant, Adrian Abbott, to be the inaugural chairman and an old contact from Sydney, Peter Bennetto, to be a director. Geoff Motteram was also on the board for his technical expertise. But in reality, Forrest, the 33-year-old wunderkind, had absolute control over his older directors and it was rare for anyone to question him.

In the early years, Forrest promised that Anaconda would only ever focus on producing nickel – a pledge that was discarded as he unveiled what appeared to be a series of hastily conceived plans to invest billions of dollars in rare earths, magnesium and phosphate projects. He also began to talk up an ambitious plan to create a "three-nickel province" around Murrin Murrin that would lead to Anaconda one day dominating global output of the metal. This predilection for hyperbole was again highlighted when Forrest declared in May 2000 that Anaconda had "discovered" an enormous inland sea underneath the desert about 600 kilometres from Murrin Murrin. The water source, known as the Officer Basin, contained 2 trillion kilolitres of potable water, enough to supply a large Australian city for many thousands of years – or so Twiggy declared.

Evoking the spirit of Sir John Forrest a century earlier, Forrest proclaimed he would green the desert. This dream won him plenty of admirers in a state where water has always been a scarce commodity. A few sceptics, however, pointed out that the state government had actually known about the Officer Basin since the 1960s and still owned the resource. Many also drew attention to the huge costs of building a pipeline from the middle of the desert into Kalgoorlie to supply water to the town and its nearby mining operations. Of course, if Forrest had been proven right, the Goldfields water pipeline from Perth may have been rendered redundant – and the history books would have recorded that he had trumped his great-great-uncle.

Forrest's grandiose public statements and his bombastic style were beginning to grate on many people. The *Australian* quoted a stockbroker who quipped that Forrest "probably feels a bit constrained by the size of the planet". The key figures along St Georges Terrace still remembered Forrest's antics in stockbroking in the 1980s and weren't about to touch his speculative nickel play, no matter how polished the sales pitch.

Forrest responded to those who doubted his credibility with a typically brilliant example of spin, which may have left some believing he'd never made an enemy in his life: "Everyone who's done business with me either respects me and likes me or just likes me. I think a lot of people who are critical either don't know me, or try to behave like they do when they don't. There's always baggage out of a boom. But there is no one who I know who can say they weren't dealt with fairly or evenly by Andrew Forrest. I have a philosophy with everyone I deal with and it's come through from my early days as a jackeroo on a sheep station, right through to my broking days, through to now: that if I win, you win. I'll never tolerate a situation where if we are both driving for the same goal, only one of us gets that goal."

Forrest's credibility was not only being challenged by some of his peers. It would soon be torn to shreds by a Perth magistrate who had closely watched Forrest's performance in the witness box during an assault case. The hearing in the Perth Magistrates Court arose after Peter Salter, the prospector who had the original Anaconda tenements, delivered a solid punch to Twiggy's head after a long lunch in September 1995. The relationship between the pair had been steadily deteriorating since their initial meeting in Sydney back in 1993. But on this day, it was to be damaged beyond repair.

Forrest met Salter at the Grand Palace Chinese restaurant in Perth's CBD to arrange for the transfer of some important tenements to Anaconda from a junior explorer called Central Bore, to which Salter was a consultant. According to testimony given in court, Forrest began telling Salter over lunch how Anaconda had tried to bring Central Bore to its knees as a way of forcing it to sell the tenements.

Salter decided to order another bottle of wine to see if any more information like this might be revealed. Forrest described in detail how Anaconda had contacted the Australian Securities Commission and the Australian Stock Exchange to raise concerns over a capital-raising Central Bore was undertaking. He also boasted of how Anaconda had tried to manipulate Central Bore's share price, sometimes using money from "Rodney's Fun Fund" – a reference to the money Rodney Adler had ploughed into the explorer. Salter also told the magistrate he had become suspicious during lunch that Anaconda was behind legal action by a mysterious Sydney-based company, Richfile Pty Ltd, which was seeking to cast doubt on Central Bore's right to an interest in the tenements.

What's not in dispute is that Forrest and Salter finished their wine and went back to Central Bore's offices to sign off on the transfer of the tenements to Anaconda for $12 million. Forrest later insisted in court he was sober at the time – but, like much of his evidence during the case, this was rebutted by others. "Always, in business lunches, I drink sparingly," he told the magistrate. "I tend to sip wine at business lunches and where possible add water to the wine glass." An employee of Central Bore, Paul Ducie, gave evidence that Forrest had trouble walking after the lunch. Even Forrest's own colleague, George Macdonald, said in the witness box that when his boss showed up at Central Bore's office, "He'd clearly had a good lunch, he'd clearly had a drink."

Forrest told the court that before the deal could be signed, Salter grew agitated and started to threaten him: "He said … I was pathetic, that he was going to get me and he was going to get my mates," he told the magistrate. Salter then punched Forrest above his left eye and again on the cheek with the back of his hand. But Twiggy didn't draw on the pugilism he had mastered in his adolescence. Instead, he calmly left the office and lodged a complaint with the police, who charged Salter with unlawful assault. Unfortunately for Forrest, the action backfired disastrously when his honesty was closely examined in court.

After listening to all the evidence, Magistrate Bob Lawrence decided Forrest had acted "unethically, immorally and at any cost" to secure ownership of the Central Bore tenements on behalf of Anaconda. He also found Forrest had provoked Salter into punching him, describing the Anaconda chief as unconvincing and a "most unreliable witness". Forrest, who wasn't represented by a lawyer in court, described the proceedings as unjust and unfair. In years to come, this became his standard response whenever a court didn't believe him.

Undaunted by his growing band of detractors, Forrest maintained a hectic pace and punishing travel schedule as Anaconda chief executive. He knew the company would need a lot more money than it had raised in the float if it was to realise its lofty ambitions. Forrest hired Peter Matheson, an engineer with vast experience working on the Greenvale nickel laterite project near Townsville. The pair toured the world, speaking to people with experience in high-pressure acid leaching and to companies who might stump up some money. In a big breakthrough in 1996, Forrest used his persuasive skills to convince the secretive Swiss commodities trader Glencore to invest $US220 million for a 40 per cent stake in the Murrin Murrin project.

The deal with Glencore appeared to be a masterstroke: Anaconda, a minnow on the global stage, would build Murrin Murrin in a joint venture with one of the world's biggest companies. As part of the agreement, Glencore would also market the nickel produced at Murrin Murrin. It was also a handy deal for Forrest personally: he earned a $1-million success fee for negotiating it, on top of his $380,000 annual salary. When asked by journalists why he was entitled to such a huge fee, Forrest snapped that he would gladly give up the money for the chance he'd lost to spend time with his two small children, after the birth of his second daughter, Sophia, the previous year. "If someone else could have achieved what has been achieved by the team, without me having to pay the monstrous penalty of not knowing my own children, because I have been totally dedicated to this company and the project, then I would have gladly paid for someone else," he said.

"I see the bonus certainly as a sign of thanks, but it's not compensation for not seeing your children grow up."

The extra $1 million would have helped pay the mortgage on Forrest's charming new house in Marmion Street, Cottesloe. The five-bedroom, three-bathroom pile, known as Burradoo, sprawled over more than 900 square metres of prime land near Perth's famous Cottesloe Beach. It was a far cry from the shared housing of Forrest's stockbroking days in Perth. He had returned home in style.

But any joy over the Glencore investment or the $1-million bonus soon evaporated when the Swiss company threatened to cut off its critical funding to Anaconda unless it could lift its 40 per cent stake in Murrin Murrin to at least 80 per cent. Forrest feared he was being squeezed out of the company he had built from scratch. "That was one of the toughest times of all of our lives involved in the company," he later recalled. Forrest realised he was now playing in the big league and would need to fight to avoid giving away control of Anaconda to Glencore, which had also accumulated about 20 per cent of the shares in the company. He was faced with a choice of selling out or rolling the dice for one last shot at survival.

True to form, he rolled the dice. In 2007, Forrest borrowed Glencore's corporate jet and flew to North America to raise $US420 million in debt from US capital markets. This erased Anaconda's short-term funding concerns and allowed construction of Murrin Murrin to begin. Forrest and his finance team had survived on a few hours' sleep for days on end to secure the debt. And like most things Forrest did, the deal ripped up the rule book. Never before had a large-scale greenfields mining project been financed by the capital markets rather than traditional bank debt. The junk bonds were interest only, which meant Anaconda would not need to repay the principal until Murrin Murrin started generating cash flow. As the *Australian*'s John McIlwraith pointed out, Forrest was making a habit of overcoming seemingly insuperable obstacles and raising money at a time when "some companies have trouble financing a new car park". Forrest attributed his Midas touch to being able to sell his vision face-to-face with the US financiers rather than dealing with a bank's lawyers and other middlemen – proof yet again that he could sell almost anything if he could just get his foot in the door.

Forrest's next fundraising mission was to London, where he took the brazen step of cold-calling Glencore's controversial founder, Marc Rich. It wasn't the first time Forrest had done business with a morally dubious character, and it wouldn't be the last. The cigar-chomping Rich had fled to Switzerland in 1983 after being charged in the United States for tax evasion, fraud and illegal trading with Iran. He had then set up his own trading business in direct competition with Glencore. Rich was a ruthless businessman who made a billion-dollar fortune doing deals with unsavoury regimes, including those of Augusto Pinochet's Chile, Nicolae Ceaușescu's Romania, Fidel Castro's Cuba and apartheid-era South Africa. He later admitted to bribing officials in countries such as Nigeria and to helping the Israeli spy agency Mossad. Rich was pardoned, in somewhat murky circumstances, by Bill Clinton on the US president's final day in office in 2001. But in 1998, when he met Forrest, the 63-year-old Rich was still on the FBI's "ten most wanted" fugitives list. Forrest was able to convince Rich to buy a 6 per cent stake in Anaconda, a move that infuriated Glencore and tempered its influence on the share register.

Back in Perth, Forrest's approach as chief executive was often highly unconventional. Before he had even raised the money from US capital markets, he began ordering the most expensive equipment Anaconda would need for the project. These included the 37-metre-long, five-metre-high autoclaves that were made by the Australian Submarine Corporation in Adelaide and trucked across the Nullarbor. Working capital was tight, so Forrest launched a competition for Anaconda staff to come up with their best cost-saving measures. One employee won a trip to Hamilton Island for suggesting that some filters be removed from the plant – a move that saved millions of dollars at the time but inevitably came back to haunt the plant's engineers. Another won a Harley Davidson motorbike for a similar idea. "They were all harebrained ideas and they all came back to bite Andrew," says a former executive.

Forrest's proclivity for living on the edge was balanced by an emerging ability to inspire intense loyalty in those working under him. This would come to be one of his foremost traits as a chief executive. Forrest was growing more confident in his own abilities as a leader and had finally lost the stutter that was still noticeable even in his early thirties. His work ethic was phenomenal. "As a CEO there's nobody I've worked with who has his energy and his enthusiasm and drive," says one veteran mining executive who worked under Forrest at Anaconda. "He would work 24-7, he expected you to do the same and he never took no for an answer." Michael Masterman, Anaconda's chief financial officer, was so loyal to Forrest's Anaconda vision that that he gave his baby son Isaac the middle name "Murrin".

By 1998, however, construction work at Murrin Murrin had started to fall behind schedule and serious technical problems were emerging. Forrest had signed a $1-billion fixed-price contract for the construction of the project with engineering firm Fluor Daniel – a deal that would ultimately prove his undoing. Experienced hands in the mining industry knew that fixed-price (or lump-sum) contracts were potentially disastrous because the project owner must cede control during construction. The contract for Murrin Murrin put the onus on Fluor, working closely with technical consultants Sherritt International, to effectively guarantee that the plant's HPAL technology would work. Any cost overruns would have to be borne by Fluor. That appeared to be a victory for Anaconda shareholders, but in reality it helped create a debacle that would ultimately lead to Forrest being ousted as chief executive.

Forrest has always claimed he had no choice but to sign the $1-billion fixed-price contract with Fluor because Glencore and the US bondholders had insisted on it as a condition of providing funding to a relative unknown in the mining industry. "We argued with Glencore and the investment banks acting for the bondholders that a lump-sum contract gave Fluor the willingness to try and cut costs from that lump sum and any cost they cut became Fluor profit," he said. "We argued that it was better for them that they left control of the construction and commissioning of that plant with management who had only one incentive and that's to make it work at the best result for shareholders."

Forrest's version of events is disputed by several insiders at Anaconda who say the chief executive was warned against a fixed-price contract. In fact, one former employee even suggests it was Forrest's exuberant brand of salesmanship that may have given the bondholders the idea to insist on such a contract in the first place.

Whatever the truth, the Fluor contract turned out to be a catastrophe for everyone involved, especially the US bondholders, who lost most of their money. Fluor couldn't deliver the plant on time and Anaconda began running out of funds. By 1999, whispers about commissioning problems at Murrin Murrin had begun to spread like wildfire through the market. The main rumours centred on problems with the giant autoclaves – the heart and lungs of the plant, through which the ore was digested in a slurry of sulphuric acid. But Forrest was having none of the gossip. He told the *Sydney Morning Herald* in February 1999: "The autoclaves are performing very, very reliably." What he didn't say, however, was that the autoclaves' flash vessels, where the acid-rich slurry was discharged, were barely working at all. When the share price collapsed 14 per cent in a single day in March 1999, the finance pages screamed that Anaconda had fallen into an "acid bath". Forrest came out swinging. He told the *Age* that although the flash vessels were not working, rumours of serious commissioning problems were groundless.

Forrest had been making forecasts of production from Murrin Murrin that were not met. Years later, Forrest would admit that the plant was seriously flawed from the start. Key equipment meant to last six months at a time, he revealed, was lasting for just six seconds.

Forrest's defining strength, then and now, is an ability to overcome hurdles that would sink most other people. Despite the diminishing market confidence in Anaconda, he was able to persuade the prime minister, John Howard, to officially open Murrin Murrin in July 1999. Howard had agreed to attend the opening because he liked Forrest, an inveterate networker who would also become close to Howard's successor, Kevin Rudd. "I've met a lot of managing directors," Howard said. "But I don't think I've met a more self-evidently enthusiastic and dedicated managing director than ... Andrew Forrest."

The formal opening was highly unusual because the Murrin Murrin plant wasn't even close to working properly. Thousands of workers had toiled on the site, in temperatures often above 45 degrees in summer, to build a project that was plagued by fatal flaws. The plant was almost incapable of producing anything at the time of the opening, says former Anaconda employee Wayne Richards. "Andrew was ringing me up saying, 'What time are the briquettes coming out of the furnace? Don't let me down, mate, don't make a fool of me,'" Richards recalls. "The briquettes only came out for two hours and then the furnace shut down. But we just hung it together for the grand opening." In fact, it would be another eleven years, with Forrest long gone and Glencore fully in charge, before Murrin Murrin would operate at its design capacity. But to those who turned up to the opening in 1999, the enormous chemical factory was impressive and imposing. With more than 3000 tonnes of structural steel rising from the desert haze, it looked like a set from a *Mad Max* movie.

The Murrin Murrin opening was also remarkable because Forrest, wiping tears from his eyes, used the occasion to lead the 300 invited guests in a lengthy and solemn prayer session. The devout Christian would later install a chapel at the remote mine site to cater for the spiritual needs of his workforce. Once prayers were out of the way, Forrest put a brave spin on the behind-the-scenes dramas. He told the audience that Fluor should be "thanked and congratulated" for its work as the construction contractor, extolling the company for "completing" the project only two years after work had begun. That would be the last time he would ever publicly praise Fluor. Just two months later, Anaconda sued the company for $1.6 billion over the construction delays and design faults. Fluor was reportedly forced to pay a fraction of this amount as part of a confidential settlement to end the claim.

In an unusual display of vulnerability, Forrest admitted that the "nightmares" of the previous two years were still haunting him. "Certainly there have been times when many believed that we would not actually get to today, and at times I shared that belief," he said at the opening. Yet despite his mounting woes, the irrepressible Forrest was still dreaming of bigger plans for Murrin Murrin. He told the audience he wanted to spend another $1 billion expanding the plant to boost production capacity by a staggering 150 per cent. By the time that was completed – as it would undoubtedly be, in Forrest's opinion – Anaconda would be the world's second-biggest nickel producer and Murrin Murrin one of the biggest mining projects in Australian history.

Warming to the task, Forrest also took a swipe at those journalists who had run stories questioning his Anaconda dream. "Where companies and individuals strive to achieve, whether in great adversity or not, they should be supported by the community, including the media," he said. "When they succeed they deserve applause. If not, they should be encouraged to try again, not vilified as a failure. It is the fear of failure, rather than the excitement of success, that drives too many of us these days, and it is this fear of failure that our press dwells on too much."

Forrest's dislike of media scrutiny did not surprise many of the journalists who had come to know him. One of them, Paul Armstrong, then a young business reporter for the *West Australian* in Perth, recalls how Forrest would regularly hurl abuse at him over stories published about Anaconda. "With Twiggy, you're either totally for him or you're the ultimate enemy – there's nothing in the middle," says Armstrong, who would go on to become editor of the newspaper and says he greatly admires what Forrest has achieved as an entrepreneur. "I dared to question whether the processing technology they had would work. He would call and say, 'Digger, what's your problem? Why can't you see there is no issue here?' It always started off with a charming introduction, and if he didn't get what he wanted, it ended with an abusive conclusion."

By the time the Sydney Olympics came around in September 2000, Forrest was slowly losing control of Anaconda. As the chairman of Athletics Australia at the time, he was among the first on the track to congratulate Cathy Freeman on winning her gold medal in the 400-metres women's final. But back in Perth, the situation was becoming dire. The Anaconda share price slumped from $4 to $1 between September 2000 and March 2001 as investors saw that the company could not deliver on the forecasts made by Forrest.

Another sin, in the eyes of some investors, was Forrest's penchant for rewarding himself handsomely even while Anaconda was losing money. His total remuneration in 2000 was $1.8 million, and by 2001 it had jumped to more than $2 million. This put him in the same league as the heads of some of Australia's biggest companies – but they were generating strong returns for shareholders. Forrest was also able to win generous performance bonuses even while Anaconda was posting heavy losses.

The company's remuneration committee was chaired by Rodney Adler, who at the same time was busy committing serious crimes in his role on the board of another troubled company, HIH Insurance. Adler would be jailed for four and a half years for a series of offences at HIH, and described by the sentencing judge as having displayed "an appalling lack of commercial morality". He quietly stepped down from the Anaconda board, but remained close to Forrest, who even visited him in prison in rural New South Wales several times.

The Olympics were a frantic period for Twiggy. As he cheered wildly for Australia's athletes on the track, he was also involved in negotiations with Melbourne mining entrepreneur Joseph Gutnick over a deal to buy his company, Centaur Mining. Forrest knew Gutnick from his stockbroking days and the two men had generally been friendly, even if some of their deals hadn't worked out. Gutnick, an ultra-Orthodox Jew, had been prominent in Australian business for years and was widely known as the president of the Melbourne Football Club in the AFL. But Gutnick's empire was laden with debt and he was desperate to offload Centaur to Anaconda. Centaur owned the Cawse nickel laterite mine, not far from Murrin Murrin, so the idea of seizing control of Centaur's assets appealed to Forrest's vision of building his "three nickel province" in the Goldfields. The Anaconda–Centaur deal was meant to have been settled in September 2000, but Forrest told

Gutnick that month he needed more time to conduct due diligence on the company.

While Gutnick waited for the deal to be finalised, Centaur's share price plummeted and the once-warm relationship between the two men turned icy. "I'm not interested in games," an angry Gutnick wrote to Forrest in November. "Centaur's future is at stake and I must act to resolve the uncertainty. You have not acted honourably, and that is very disturbing to me." A day later, Anaconda told the market it was backing out of the deal. Gutnick responded with legal action in the Victorian Supreme Court, claiming Anaconda had breached the agreement by failing to complete the transaction. Forrest appeared to be affronted by the litigation. "There has not been an obligation that Anaconda hasn't honoured and we're not about to break that record with Joe Gutnick," he said.

Justice Marilyn Warren's findings against Forrest in 2001 were damning. She ruled that he had pursued a "deliberate plan" of maintaining the pretence of a share deal with Gutnick so he could prevent a rival to Anaconda from gaining a foothold in the "three nickel province". The delay, she found, was also aimed at allowing Anaconda to gain control of Centaur's nickel assets on the cheap. She said Gutnick was probably foolish to have trusted Forrest in the first place. The "smoking gun" in the case was an email written by Forrest's right-hand man at Anaconda, Stephen Dennis. In the email, Dennis told Forrest he should falsely explain to Gutnick that Anaconda was backing out of the deal "because of due diligence reasons". It was advice Forrest took on board. Justice Warren took a dim view of the email. "The Dennis memorandum set out a strategy to make Gutnick dependent upon and vulnerable towards Anaconda, gain control of Centaur and quarantine Gutnick from Anaconda's rivals," she found.

But then came the killer blow. In assessing the details of an important phone call between Forrest and Gutnick, the judge decided that she believed the latter's version of events. "Gutnick was a truthful witness in all respects of his evidence," she said. "Having observed Forrest, including his physical demeanour, especially in the course of giving evidence-in-chief, I am unable to accept his version of the conversation of 4 September 2000. Forrest was an untruthful witness."

As he had done previously, Forrest went on the warpath over the 110-page judgment. "It's a strange judgment … it's most extraordinary and logically unbelievable," he said. "It's just a really dumb outcome." Forrest said the judge had been persuaded by a "very clever Queen's counsel" acting for Gutnick. And on Justice Warren's key finding that he had not told the truth in court, he said: "Anyone who knows me knows my great character." Ever the optimist, Twiggy expressed his absolute confidence that an appeal court would overturn Justice Warren's verdict. But neither the Victorian Court of Appeal nor, ultimately, the High Court could see his side of the story and the decision stood.

Two others who had long felt let down by Andrew Forrest were Canadian investment bankers Stephen van der Sluys and Richard Maish, who had helped Anaconda negotiate the $US420 million in junk bonds from US capital markets in 1997. Van der Sluys and Maish, who were employed by the Canadian bank CIBC, sued Anaconda in the NSW Supreme Court, claiming they had not been paid a success fee to which they were entitled. A critical part of the case revolved around the details of a phone conversation between Forrest and a senior executive at CIBC, Bruce Spohler.

When Justice John Brownie came to rule on the issue in 2002, he was in no doubt as to whose evidence he preferred about the details of the phone call. "I prefer the evidence of Mr Spohler who I regard as a truthful and reliable witness, to that of Mr Forrest, who I regard as quite untruthful," he said. "Indeed, I think it would be unsafe to rely on any account he has given, in or out of court, except to the extent that it is demonstrated by other evidence to be correct. His evidence was generally quite unimpressive, and his evidence about the telephone conversation of 7 August 1997 particularly so."

For the fourth time in his career, in four unrelated legal cases, Forrest had been found to be a less-than-honest person. In all four cases, his ethics and truthfulness had been impugned. It is an extraordinary record that is unusual for someone who has run a large listed company in Australia.

Forrest has complained that other courts have found him to be an honest witness, but that this has never been reported. The sole reference to his honesty, however, was a 1999 case in the WA Supreme Court in which Justice Eric Heenan found that Forrest's version of two phone conversations with a fellow mining company executive, Steven Dean, was "both truthful and substantially accurate". The judge also noted that Dean did not give evidence in court over the dispute with Forrest, and he ended up ruling against Anaconda in the case.

Many of Twiggy's other business disputes were settled before a judge could rule on them. One involved a falling out between Forrest and two of the central figures at Anaconda, Peter Matheson and Geoff Motteram. In the early days of the project, the pair formed a company, Matheson Motteram Projects, to advise Anaconda on how to build the nickel processing plant. They had also negotiated with Forrest to be paid 0.5 per cent of the capital cost of Murrin Murrin, assuming it was ever built. But years later they alleged Forrest had no intention of paying up and sued Anaconda in the Supreme Court of Western Australia. The case was eventually settled, with Matheson and Motteram receiving about one-third of what they claimed to be owed.

Geoff Motteram had originally made his feelings known about Forrest at Anaconda's fancy-dress Christmas party in 1997.

As scores of the people present at the bash at Perth's Burswood Casino can attest, the metallurgist had downed a few too many beers during the evening and began loudly criticising Forrest to then Anaconda chairman Allan Coogan. Forrest, who was dressed as Elvis Presley in flares and sideburns, was not within earshot at the time. But the chief executive soon heard about Motteram's rant. Two days after the party, Forrest turned up at Motteram's house in Dalkeith with a cardboard box filled with the contents of his office. He told him he was no longer a director of Anaconda.

The bad blood between Motteram and Forrest didn't end there. In 1999, Motteram was among the hundreds of people invited to the formal opening of Murrin Murrin. But in the weeks leading up to the event he had sold a large number of Anaconda shares. Forrest sent Motteram a fax uninviting him from the opening, saying only people committed to supporting the company were welcome to attend. The fax bemused Motteram, given that he still held plenty of shares. But Forrest could see nothing odd about it. In his eyes, anyone who sold shares was not a true believer and therefore did not deserve to be at the event. "The numbers were tight," Forrest told the *West Australian*. "Those who sold shares have had their invitations withdrawn and replaced by those who bought shares."

Forrest's fate at the helm of Anaconda was effectively sealed just a few weeks after Murrin Murrin's formal opening when he persuaded South African mining giant Anglo American to invest $US243 million for a 23 per cent stake in the company, putting it above Glencore at the top of the share register. At first blush, Anglo appeared to be Forrest's saviour. The company had agreed to buy shares in Anaconda for $3.15 each – a premium of more than 35 per cent on the market price. In one fell swoop, the world's second-largest mining company had solved Anaconda's financial troubles and given it a huge credibility boost in the eyes of the market. Within months, Glencore's sole director, Michael O'Keeffe, had been voted off the board with the backing of Forrest, Marc Rich and Sherritt International, which owned 9 per cent of the stock. Twiggy's old business partner, Albert Wong, looked on in admiration at Forrest's "divide and conquer" tactics, saying the idea had probably germinated when the pair first read Chinese military strategist Sun Tzu's *The Art of War* a decade earlier.

Forrest, with his back against the wall, had employed his trademark salesmanship to convince Anglo's hard-headed minerals boss, James Campbell, that Anaconda Nickel, despite all its woes, still had the potential to dominate the global nickel industry. Eventually, however, Campbell came to believe that Forrest would have to be removed if Anaconda were to realise that potential. In 2001, with Murrin Murrin still in deep financial and operational trouble, Anglo declared it had lost all confidence in Forrest's management. It told investors that the company's "missed targets, lost production, lax corporate governance, capital cost overruns and high-risk capital structure" had cost Anaconda shareholders up to $431 million in lost profits. Anglo called a special shareholders' meeting in an attempt to sack Forrest as chief executive, along with most of the board, promising to recapitalise Anaconda with a $100-million rights issue. Forrest denied the allegations about mismanagement and claimed Anglo was simply engaged in a ruse to try to seize control of Anaconda on the cheap.

Drawing on all his street-fighting instincts, Forrest refused to bow to the might of Anglo American. He boarded a plane and went to Europe in search of another potential white knight who might save him. But Twiggy's message by now was wearing thin. More than two years after John Howard had "opened" Murrin Murrin, the plant was still not producing the nickel Forrest had so confidently predicted, at the low costs he'd always promised. Anaconda was saddled with debts of more than $800 million and wasn't even close to turning a profit. In what amounted to a high-stakes game of corporate chess, a stand-off ensued for several months. Nobody appeared to know whether the enigmatic Glencore, which held the balance of power, would back Forrest or Anglo American.

At an eight-hour meeting in Perth in May 2001, the warring parties – Forrest, Glencore and Anglo – thrashed out a compromise under which Twiggy would stand down as chief executive in November but remain on the board as deputy chairman. Forrest's dream was now over. He was deeply wounded at being ousted as chief executive of the company he'd built from nothing, but as always, he attempted to put a positive spin on the move and laced it with some overblown rhetoric. "The winner is an independent Anaconda Nickel," he declared after the meeting. "I think what has been achieved is an independent Australian company now fully supported by major shareholders, as opposed to dominated by them, and I think that is an excellent outcome for corporate governance and democracy in this country."

Forrest left Anaconda having achieved one of the most extraordinary fundraising operations even seen in corporate Australia. In total, he had convinced investors to part with almost $US1 billion – or about $1.4 billion in Australian currency terms – purely on the strength of his fanatical conviction that Anaconda was the next big thing.

Forrest does not deserve to shoulder all of the blame for Anaconda's woes. Fluor was responsible for many of the design flaws and delays that cost the project dearly. Forrest's Anaconda vision has also been vindicated by the belated success of Murrin Murrin, now fully owned by Glencore, in recent years. Some of the brickbats that have been thrown at him since 2001 are simply a result of the old adage that history is written by the winning side. But his plan for Anaconda was too big and too hurried. His failure to take advice from some of the experts around him was also an error.

Says one former Anaconda executive: "Andrew knew it all – that was his biggest problem. He didn't have a clue [about the technical side] ... he should have taken more advice."

For his part, Forrest has never publicly accepted any of the blame for what happened at Anaconda, preferring to cast himself as the victim of forces motivated by greed. "It taught me cynically that the bigger the numbers become, the more powerful are the interests to act indecently or act purely for business and not for moral or social responsibility, or for a grounding of social values," he said in 2012. In private, however, Forrest has told friends that his biggest mistake at Anaconda was not ensuring that he was briefed about the design changes made along the way.

Anglo American sold out of its disastrous Anaconda investment in 2003, leaving Glencore to take full control, with the aim of spending even more money to rebuild the company from the ground up. A bitter Forrest, who still owned a 4.5 per cent stake in Anaconda, would later back a hostile $450-million takeover attempt by US vulture fund MatlinPatterson, arguing it was needed on the grounds that Glencore's capital raisings were diluting other shareholders. The bid failed and Forrest sold all his remaining shares.

A former Western Mining Corporation executive, Peter Johnston, succeeded Forrest as chief executive and began to put his mark on the company, which was in 2003 renamed Minara Resources in a bid to disassociate itself from the Forrest era. Johnston soon ditched Anaconda's much-hyped "three nickel province" strategy and focused on making the Murrin Murrin plant work. The company was effectively insolvent and Johnston had to fly to New York to convince 100 bondholders owed $US400 million to take a big haircut. In the end, they received less than 25 cents in the dollar on their investment. Johnston says Forrest deserves credit for driving Anaconda as far as he did. "There's no doubt it wouldn't have happened without him," says Johnston, who took more than ten years to fully fix the problems he inherited in 2001. "But basically he was an entrepreneur, and I think fundamentally he didn't understand the technology. I think his lack of experience in the end told. But boy, was he an entrepreneur!"

More than 100 Anaconda staff with links to the Forrest era were dismissed at the end of 2001. But even as he walked out of the Anaconda office for the final time, Forrest maintained a veneer of positivity. Recalls one former employee: "I remember Andrew coming through the office and saying, 'Guys, I'm off, Peter Johnston is a terrific fella, he'll look after you.' Of course, that afternoon we were all retrenched." The former employee suggests he and his colleagues felt their years of intense work and loyalty to Forrest were not fully repaid. "Andrew didn't even have a beer with us. He asked everyone to sell their souls and give up everything for him, and yet he sailed off in reasonable shape. Did he know we were all going to be retrenched? I suspect he did. He demands great loyalty and I'm not sure he gives as much back."

What nobody realised at the time, however, was that Forrest was not in reasonable shape. In fact, he was facing a personal financial crisis in 2001 that at one stage threatened his family home in John Street, Cottesloe, a historic mansion set on an 1860-square-metre block that he and Nicola had bought a year earlier, around the time that their son, Sydney, was born. Forrest had borrowed heavily to buy most of his 32 million shares in Anaconda and had watched in horror as the price plunged, triggering margin calls from his stockbroker. Forrest has never publicly commented on his brush with financial ruin. But in a statement lodged with the Administrative Appeals Tribunal in 2007 as part of an unrelated legal case, he admitted he had been under "financial pressure" from his bank in September 2001 because he had borrowed "substantial amounts of money" to buy Anaconda shares.

Forrest sought advice from several people on how he might rustle up a few million dollars quickly and was told to get on a plane to see Shayne Heffernan, a shadowy Australian expatriate who had become prominent in Asia as a high-risk venture capitalist. Heffernan, who stands about 190 centimetres tall and is heavily built, appeared to be making serious money and was living in the penthouse suite of the Shangri-La Hotel in Hong Kong, with a chauffeur on call to ferry him around the city. But he was never in one place for long and would be forever darting around the region.

Heffernan had a colourful past. He had been arrested in Sydney in 1987 and charged with supplying heroin as part of a "sophisticated network" that had allegedly sold more than $5 million worth of the drug. The Central Court was told that Heffernan sold the heroin at street level and had also investigated an abortive mission to take $40,000 cash to South-East Asia in a "narcotics-related exercise". After a ten-month trial in the late 1990s, he was found not guilty by a jury of selling heroin, but was convicted on the lesser charge of supplying marijuana and imprisoned.

Heffernan next came to the attention of the Australian authorities in 2003, by which time he had moved to Asia and reinvented himself. The Australian Securities and Investments Commission issued a statement warning that hundreds of local investors had transferred $3.5 million to Heffernan's Hong Kong-based company, Equity-1, on the promise that they would receive returns of 30 per cent through international share trading. The Hong Kong Securities and Futures Commission claimed Equity-1 was involved in "boiler room" activities, a fraud involving high-pressure sales tactics.

Nobody really knew how Heffernan had made his millions, but Forrest wasn't about to ask too many questions. On the night of 11 September 2001, he was in Heffernan's suite at the Peninsula Hotel in Manila, watching the television as planes slammed into the World Trade Center in New York. Heffernan says he believes he and Twiggy bonded that night, and he agreed to lend his new mate $3 million so he could pay off his debts and hang on to his house. "I called the guy at the bank to make sure it was real – and he said he was going to take possession of the house and his credit cards were going to be cut off," Heffernan recalls. "Forrest told me: 'When I have cash in bank, you're first on the list, I'll never forget you, mate.'"

By early 2002, Forrest's financial pressures had eased and he embarked on a twelve-month holiday to Europe with Nicola and their three children, Grace, Sophia and Sydney, who were all younger than eight. After promising to remain on the Anaconda board, he suddenly quit as deputy chairman in March and flew out of Perth the next day. Forrest's reputation was in tatters and his spell in corporate exile had begun.

# 7.

# THE COMEBACK KID

*To go back into the limelight after he'd been
vilified like that, I think was incredibly brave.*
—NICOLA FORREST

It's just after 10am on Friday 18 July 2003 when Andrew Forrest rises to speak. His audience is a small group of mainly elderly shareholders who have gathered on the second floor of the 100-year-old Celtic Club in West Perth to hand him control of their obscure company, Allied Mining and Processing. They know Twiggy has a reputation as a corporate cowboy. But by installing him as their chairman and major shareholder, they hope he can develop Allied Mining's old iron ore tenements in the Pilbara – and perhaps even make them some money. With iron ore trading at a paltry $US30 a tonne, it might not be a lot of money, but it's worth a punt. Few of the shareholders listen too closely to Forrest's bullish presentation. As soon as he's finished, they hurry downstairs for coffee and biscuits before heading back outside into the morning chill.

Those who did pay close attention to Forrest on that day in 2003 would have heard one of the more prophetic speeches in Australian corporate history. Within a few minutes, he had accurately predicted the imminent arrival of the greatest mining boom Australia had ever seen. Displaying a series of simple slides, Forrest declared he would transform Allied from a penny-dreadful stock into the "third force" of the Pilbara iron ore industry, directly challenging the stranglehold enjoyed by mining leviathans BHP Billiton and Rio Tinto. He told the audience that global demand for iron ore was set to rise from 480 million tonnes a year to 700 million tonnes a year within just four years, but that existing miners would only be producing 620 million tonnes by then – a global shortfall of 80 million tonnes. Allied Mining and Processing, he explained, would fill the gap. Some may have smirked, but Forrest seemed to believe what he was saying.

As he had done at Anaconda Nickel, Forrest made it all sound so easy. He would build a mine, a railway and a port to sell iron ore to the steel mills of China – an economy, he forecast, which would grow rapidly for decades to come. The whole thing would cost about $2 billion, which he would raise on world markets. One of the slides that flashed up on the screen audaciously spelled out his main goal: "To create an Australian-owned iron ore group comparable in size with the current majors."

The company would soon change its name to Fortescue Metals Group, after the Fortescue River, which snakes its way through the Pilbara heartland. Reflecting on that first meeting a few years later, Forrest said: "I thought, [it] will be like Napoleon said: let China sleep, because when she awakes she will shake the world. We took the view in 2003 that China had awakened." Others were making similar predictions at the time, but Forrest was among the first to stake his career on the hypothesis that the rise of China would be a multi-generational phenomenon rather than an economic blip.

Three months before the Allied shareholders' meeting in West Perth, Forrest was sitting at home in Cottesloe mulling over his plan to create a new mine in his native soil of the Pilbara. He had been in exile for almost a year and, still aged only forty-one, he had the itch for a new adventure. Other Perth entrepreneurs had dreamed for decades of the iron ore wealth on offer in the Pilbara – including Alan Bond, who in the 1970s had a big stake in mining company Robe River before it was taken over by Rio. Most were deterred, however, by the fact that the best deposits had been cherry-picked by BHP and Rio, who also owned the railways needed to carry the ore hundreds of kilometres to the coast.

Forrest's quixotic vision really began to take hold when an old friend doing drilling work for Rio revealed to him that the mining giant was doing virtually no fresh exploration work in the vast region. Forrest realised there could be untold riches on offer. But with his reputation at rock bottom, he had few friends in the mining game he could call on for advice. He only had to read the business pages of the newspaper each day to be reminded of the financial disaster he had left behind at Anaconda Nickel.

Forrest did have one old mentor he could turn to: Graeme Rowley, a 63-year-old former air force pilot who had retired after working as a senior executive at Rio's rail and port operations in the Pilbara. Forrest and Rowley had met when Twiggy dated his daughter, Lisa, in the 1980s. The avuncular Rowley happened to be driving through Cottesloe when he took Forrest's call. "Andrew said, 'Come and have a scotch, I've got an idea,'" recalled Rowley, who arrived at Forrest's house in John Street within minutes. Over a long chat at the kitchen table, Forrest outlined his belief that the Pilbara was chronically under-explored and told Rowley of his plan to take control of Allied Mining, which had a vaguely promising iron ore deposit called Mount Nicholas, just north of the Fortescue River. "I laid out a map of the Pilbara and I drew on it with texta where the railway line was going to go and where the port was going to go and where I thought we could find ore points. And he [Rowley] said, 'How much do you think this is going to cost?' And I said, 'Between 2 and 3 billion.' Now the market value [of Allied] was about $6 million. He just shook his head and he said, 'Does Nicola know anything about this?' I said, 'Yeah, she thinks I'm nuts, but we're going to have a crack.'"

Rowley was so taken with Forrest's enthusiasm that he agreed to commit his entire personal savings of $80,000 towards buying shares in the new venture, although only after Twiggy had told him he liked his executives to have "skin in the game". It would be the best financial decision Rowley ever made. He agreed to join Forrest as his first employee, a move that gave much-needed credibility to the new enterprise. "After that first meeting with Andrew, I knew for the first time in my life I would be putting my reputation on the line," Rowley said. "But I had realised what I had been looking for was the buzz that happens when you are exposed."

Rowley, Forrest and a personal assistant, Louise O'Reilly, moved in to the home's study, with its towering bookshelves, leather armchairs and Forrest family memorabilia. Next they hired Alan Watling, who had built heavy-haulage railways for Rio. Another experienced mining executive, Chris Catlow, was brought in as chief financial officer. But Catlow, like others in those days, hesitated before taking the job. "A few friends of mine were very questioning when I told them that I'd taken on the job," he recalled. "They said, 'Are you really sure you want to associate yourself with Andrew Forrest – he's on the nose.' I guess that was the common view, that Andrew was a little wild, if not a cowboy."

Forrest then attempted to recruit John Hancock, the only son of Perth iron ore heiress Gina Rinehart. Hancock, who had a strained relationship with his mother, had left Hancock Prospecting, the family company that had been established by his legendary grandfather Lang. "That first meeting with Andrew reminded me of the grand visions my grandfather would discuss with me when I was a child," says John. "It was familiar for me to hear about yet-to-be built railways, mines and ports – feats of engineering on a massive scale." Forrest offered Hancock, who had experience in dealing with Chinese steel mills while at Hancock Prospecting, a job as his new company's head of marketing. He also wanted to throw in a big pile of share options.

John put the idea to his mother, who rejected it out of hand. "I told her Andrew was starting his own iron ore company regardless, and that Hancock should get involved at this early stage – if we had a decent equity position I would be working in Hancock's interests rather than in competition. She did not want to get involved. With hindsight, it would've been an outstanding investment. But my mother was by no means alone in the mining industry in viewing this new entrant with suspicion."

John Hancock next proposed to his mother that Hancock Prospecting sell some of its Pilbara tenements to Forrest in return for a major stake in the new venture. In effect, he was proposing that the two iron ore explorers should work together against the might of Rio and BHP to develop mines and infrastructure. The plan could have shaken up the Pilbara mining industry, but it was also rejected by the steely Rinehart. More than a decade later, Rinehart and Forrest have never cooperated on anything and still view each other with suspicion. In fact, Rinehart tried unsuccessfully in 2007 to block the granting of five mining leases to Fortescue on her Mulga Downs pastoral station. Never previously known as an environmentalist, she lobbied the government to stop Fortescue because of the apparent threat that mining would pose to the endangered night parrot on her land.

Forrest, as Fortescue chairman, assembled a board of directors that drew heavily on the contacts he'd made since 1999 as chairman of Athletics Australia, where he had been enlisted to provide the sport's governing body with some business acumen. British athletics legend and Conservative Party identity Sebastian Coe was an inaugural director of Fortescue in 2003, although he resigned the following year to work on London's ultimately successful bid for the 2012 Olympic Games. Coe handed back his 750,000 share options in the company when he resigned, missing out on a multi-million-dollar payday he'd have been eligible to receive at the height of the company's success several years later. Herb Elliott, the celebrated Perth-born athletics champion who had forged a successful business career in Sydney, was recruited as a director. Another Sydney businessman whom Forrest had met through athletics, a former executive of Optus and the Commonwealth Bank, Russell Scrimshaw, also agreed to join the board. The other two original Fortescue directors, Ken Ambrecht and Chris Linegar, brought strong international finance connections.

Apart from Forrest and his cohorts, few people were excited about Fortescue's prospects in those early days. The company's share price remained at rock-bottom levels well after the July 2003 meeting that appointed Forrest as chairman. Where Twiggy had once been treated like a rock star at the annual Diggers and Dealers mining conference in Kalgoorlie, hardly anyone now wanted to be seen with him. One night, a desperate Forrest ended up drinking with a few journalists at the conference – and even they told him bluntly what a disaster he'd created at Anaconda.

He later described the campaign against him at the time as "corporate cruelty", but said it taught him to be more resilient. "When you are feeling things against you – pain, unfairness and sorrow – that is when character grows," he said. Nicola described her husband's decision to establish Fortescue after being forced out of Anaconda as "the bravest thing he's done". "To go back into the limelight after he'd been vilified like that, I think was incredibly brave," she said.

It was both brave and incredibly clever. What few realised at the time was that Forrest had negotiated a brilliant deal with Allied Mining boss Mark Caruso that would make him a billionaire several times over when investors finally clicked to his plan.

Caruso approached Forrest in early 2003 because Allied was sitting on nearly a billion tonnes of iron ore and needed someone with the energy to develop the deposit. "I had seen what he had done at Anaconda and I knew he could sell ice to the Eskimos," Caruso says.

Forrest acquired his controlling stake in Fortescue for a meagre 8 cents a share – the vesting price of more than 100 million options he was granted in 2003 under the deal with Caruso. He was also given 5 million shares for free. By January 2005, Forrest had converted all of his options, paying only $8 million for a 47 per cent stake in Fortescue. He described the decision to buy the shares as a "remarkably selfless corporate act" because it had injected capital into Fortescue. Just how selfless is a matter of debate. For by this time, the Fortescue share price had soared to $4, thanks largely to Forrest's promotional abilities. His $8-million investment was already worth more than $400 million. It was the deal of a lifetime – and it only got better. In February 2005, just two weeks after converting the last of the options into shares, Forrest sold 3 million of them for $4.50 each, pocketing $13.5 million. The $8 million he had paid for his shares was already a distant memory: he was ahead by $5.5 million on the deal, and he still had 102 million shares.

Forrest had pulled off a deal that within a few years would make him Australia's wealthiest person. But he had done it without, as he had exhorted Graeme Rowley and others to do, putting any of his own "skin in the game". In spite of this, Forrest has constantly stressed the huge gamble he took to achieve his dream. "Fortescue was started by a few friends with big ideas sitting at a kitchen table," he told the National Press Club in Canberra in 2012. "It was a huge risk. The very real possibility of failure was our constant companion."

Forrest's motivation in negotiating the deal with Caruso was not simply to get rich fast. He had learned from bitter experience at

Anaconda that he would need to be the dominant shareholder in Fortescue to call the shots and fend off potential predators. These days, Mark Caruso sees himself as the man who helped make Forrest a household name. As he relaxes on the balcony of his Perth CBD penthouse apartment, it's clear that Caruso, who is in his early fifties and talks with machine-gun speed, has also done very well out of the mining boom. But few people realise that it was Caruso's decision to approach Forrest about the Allied tenements that changed the course of mining history. "Andrew hugs me when he sees me, because he knows the true story," Caruso laughs.

Caruso, however, rejects suggestions that Allied gave away the shares to Forrest too cheaply. "Iron ore back then was sewerage shit – who wanted it?" he says. "If you didn't own a railway track, iron ore was no good to you." In fact, Caruso says, Forrest wanted a lot more shares than the 110 million he ended up with. He recalls: "Negotiations were colourful, it's fair to say. I went to his house a couple of times and in the end he just wore me down. Looking back, did I do a bad deal for shareholders? No, I did not. If I didn't give the man an incentive he was never going to do it."

Caruso does admit, however, that Forrest has an almost mystical ability to convince people to do things. In 2005, Caruso was in South Africa when he received a call on his mobile phone from Forrest. It was the same day that the 1 million options Caruso held in Fortescue could be exercised. Forrest, who wanted the money in the company, asked Caruso to convert the options into shares for $200,000. "The bank wouldn't transfer the money because they wanted my original signature so I rang Andrew and said, 'I can't do it, mate.' Forrest then gave him a motivational talk. "So I rang the bank back and yelled, 'You'd better transfer that fucking money!' He wound me up to that point. It's hilarious how he has that power, it's like a magnetic enthusiasm that draws you in. It's extraordinary, I've not seen a man like that."

Forrest's next move was to hire another former Rio executive, Philip Kirchlechner, as Fortescue's head of marketing, a role that would entail selling iron ore to China's steel mills before it had even been discovered. Kirchlechner, who had worked in Shanghai, was astounded by Forrest's energy when he first met him in his Cottesloe study. "It felt like the movie *Wall Street*, where Charlie Sheen meets Gordon Gecko for the first time and he walks out mesmerised by what he's just heard," he says. "As I was walking out I could hear him say, 'Get me the prime minister's assistant on the phone, please.' At the time I didn't quite believe it. But later he told me that whenever John Howard was in town, Howard would contact him in some way or another."

Forrest was determined to move at a breakneck pace. In August 2003, he outlined plans to start building Fortescue's railways by early 2004 and to ship iron ore to China by 2006. The target for first exports was then delayed until 2007, which later became 2008 – still an ambitious deadline, given Fortescue hadn't even existed until mid 2003. It was Forrest's ability to inspire others to work towards these unrealistic goals that would become his trademark at Fortescue. As he had done at Anaconda, Forrest dismissed the unbelievers who raised doubts about technical problems or cash shortages. His self-promotional abilities also had not waned. A photograph on the cover of a glossy Fortescue brochure in the early days showed a beaming Forrest surrounded by prime minister John Howard and senior ministers of the Coalition government. To the Chinese investors and customers who saw it, Forrest appeared to be at the centre of national power.

At Forrest's house in John Street, Cottesloe, the number of people turning up to work each day continued to grow. The Fortescue team soon moved out of Forrest's study and into the larger dining room. After a few weeks they also took over the family's living room to create a bigger command centre. Nicola, who was looking after three young children, was relieved when her husband and his new colleagues finally found an office in the leafy suburb of Nedlands in October.

When geologist David Mendelawitz started at Fortescue in 2003, he was convinced he'd made a bad decision. Mendelawitz was hired in the same week as another geologist, Eamon Hannon, and the pair would become key figures in Fortescue's success. They headed straight to Fortescue's main mining tenements at Mount Nicholas in the Chichester Range, but soon realised the low-grade ore was riddled with shale and other impurities and would never be economic to mine. The company's exploration campaign also seemed poorly targeted and unprofessional. For Mendelawitz, it appeared Fortescue was just another small explorer with outsized dreams. "It was the most disorganised show I'd ever seen in my life," he says. "Eamon and I said: 'Should we leave today or should we see it out until bankruptcy?'"

A Fortescue metallurgist, John Clout, had the unenviable task of telling Forrest about the woeful drilling results at the Mount Nicholas tenements, which Allied Mining had once estimated contained up to 1 billion tonnes of iron ore. It looked as though Forrest had been sold a pup. At this point, the Fortescue dream could easily have ended. Some of the company's geologists began openly flicking through newspapers in the office, hunting for other jobs. But Clout, who had worked for the CSIRO and is regarded as one of Australia's foremost experts on iron ores, had an idea. He ran some tests on the samples from Mount Nicholas and was surprised to find very small amounts of what is known as microplaty haematite – the same high-grade ore that is mined in huge volumes by BHP and Rio in the Hamersley Range to the south of Mount Nicholas. Microplaty haematites had never been detected in the Chichester Range and Clout was ridiculed by some for even suggesting it existed there. But he was convinced he was right; the challenge would be finding enough of the premium ore to justify a mine.

Until that time, the conventional wisdom at BHP and Rio was that potential ore bodies in the Pilbara ended only a few hundred metres from any iron ore outcrop. But John Clout and David Mendelawitz suggested Fortescue's geologists should drill further out, in the hope of finding thinner "sheets" of iron ore over an extensive area rather than the usual deep deposits. It was another example of how those at Fortescue, spurred on by Forrest's leadership style, saw themselves as challenging the status quo. "We had to think outside the box," says Mendelawitz. "BHP and Rio didn't have any entrepreneurial people – they would have to go through a lot of processes and procedures to get anything done. We wrote the procedures after we did them."

One of the region's more promising deposits, Christmas Creek, was located on the 400,000-hectare Roy Hill pastoral station, whose grizzled owners, Ray and Murray Kennedy, wore pistols on their hips and hated the mining industry. Fortescue did not have any tenements covering Roy Hill, but Forrest was able to sweet-talk his way on to the property using his old pastoral industry connections. He told the Kennedy brothers that he would pay them if they drilled some bore holes on their land. If they found water, they could have it – as long as they handed over the dirt to him for testing. The brothers agreed and the results vindicated the theory that the Chichester Range ore stretched for kilometres in shallow deposits. "We assayed the samples," Forrest recalled, "and we picked up huge traces of [microplaty haematite] again and that's when we thought we have a sheet on our hands, a huge sheet of iron ore."

Forrest put Eamon Hannon in charge of exploration and the pair became close. Former executives say Forrest and Hannon got on so well because both dared to dream about the Pilbara's potential. "Andrew liked Eamon because he used to promise the world; he treated him like a son," says one former employee.

David Mendelawitz says Hannon was the pivotal figure in Fortescue's exploration success between 2003 and 2005, when the company spent $170 million drilling an astonishing 9000 holes. "Morale was good with Eamon in charge," he says. "He would take a bullet for his team and he would stand up to the executives in Perth in the name of employee welfare." Hannon also developed a reputation as a party boy. A keen surfer, he recruited old mates from around his home town near the surfing mecca of Margaret River to work in his exploration team in the Pilbara. After a scorching day's work drilling in the dust, plenty of alcohol flowed freely each night among the exploration team, most of whom were in their twenties and thirties. The workers even created a makeshift nightclub, dubbed Club Strange, by pushing together two old caravans, fastening speakers to the walls and outlining a dance floor on the red dirt.

With Fortescue's cash reserves running perilously low in mid 2004, Forrest asked Hannon to see if he could find iron ore closer to Fortescue's proposed port at Port Hedland, in an attempt to reduce the rail costs of the project. By this time, Hannon had been instilled with the Forrest-led mantra that "nothing is impossible". Those who doubted they would find the ore in the middle of nowhere were banished from the exploration camp. "I wouldn't have anyone up there who didn't assume we were going to be successful," Hannon said. "Come hell or high water, we were going to find the ore somewhere. It was just non-negotiable. And everyone was empowered with that vision."

On 9 June 2004, Hannon made a monster discovery of iron ore to the west of Christmas Creek that would become known as Cloudbreak, named after a famous hollow left-hand curve in the waters of Fiji. At last, Fortescue had enough iron ore for a mining project. In fact, it had enough to last decades. An emboldened

Forrest then told Hannon: "Scour the entire Pilbara and don't stop until you find 10 billion tonnes of iron ore." Forrest decided that Cloudbreak would be Fortescue's first mine, followed by Christmas Creek. The two deposits have 1.5 billion tonnes of iron ore between them.

A year later, Fortescue geologists Neil Clarke and Doug Kepert were able to decipher clues from forty-year-old geological maps to discover another rich deposit called Solomon, containing 3 billion tonnes of iron ore. Unlike the earlier discoveries, Solomon was smack bang in the middle of the rust-brown hills of the Hamersley Range, which had been the domain of BHP and Rio for forty years. In fact, Rio had once held the Solomon tenements but discarded them as surplus to requirements. The Solomon find was doubly pleasing for Forrest because the Hamersley Range was named in 1861 after the family of his great-great-aunt, Margaret Hamersley, the wife of Sir John Forrest. Margaret's father, Edward Hamersley, a wealthy landowner, had sponsored the first expedition to the area. Andrew Forrest would later compare the work of his ancestors in "opening up" the Pilbara to Fortescue's "modern-day explorers and developers".

The discoveries vindicated Forrest's aggressive and unprecedented campaign to acquire huge numbers of mining tenements. The company began life with tenement holdings in the Pilbara covering only 487 square kilometres. But a small team of employees led by geologist Damon Edwards sat glued to computer screens all day, instantly picking up any ground that became available.

Most of the tenements had been previously explored by BHP and Rio but discarded because they contained what was believed to be poor-grade iron ore. In some cases, the big miners simply hadn't drilled enough holes to realise the extent of the mineralisation that existed under their feet. Forrest ordered his team to peg every inch of land they could, regardless of past geological findings. It was to prove the making of the company. By the end of 2004, Fortescue's tenements covered almost 4000 square kilometres. Four years later, when the miner began exporting iron ore, it had 52,000 square kilometres, making it a bigger tenement holder than either BHP or Rio. Today, it has more land than BHP and Rio combined, with tenements covering an area more than 1500 times larger than Sydney Harbour. At one point, Rio's then head of iron ore, Sam Walsh, referred to Fortescue's rapidly growing land bank by quipping that it appeared to be operating in real estate rather than mining.

After acquiring so many tenements, Forrest thought he had a strong case to put to BHP and Rio for them to share their infrastructure in the Pilbara. It would be pointless, he argued, to duplicate the region's existing railways. He put a proposal to the majors centred on improving the operating efficiencies of their rail and port operations. But the big boys viewed Forrest as an irrelevant minion who was attempting to gain a free ride on the railways they had built and operated for forty years. Forrest's meetings with BHP and Rio didn't last long, but they did serve to bolster his belief that the Pilbara was a kind of El Dorado. "The rejection I received was so blunt and so firm that it really firmed my resolve that there must be something really very interesting there, if these companies would deliberately choose inefficiency in order to keep ports and rails as barriers to entry," he said.

When Herb Elliott pleaded Fortescue's case during a meeting with then BHP chairman Don Argus in 2003, he too came away with nothing. Elliott recalled: "I said to [Argus] that we can both save a lot of money by sharing facilities and working together, and he said, 'Quite frankly, I've never heard of your company but leave it with me, I'll talk to the staff about it and I'll give you a ring in a couple of weeks, Herb.' Well, that's ten years ago and I still haven't heard from him."

Forrest didn't abandon his plan to force open the BHP–Rio railways. In 2004, Fortescue applied to the National Competition Council to "declare" one of BHP's railways as critical national infrastructure, which would force the mining giant to carry its ore. Wrapping himself in the flag, Forrest argued that his application to the NCC was in the national interest as it would create opportunities for all small explorers with stranded deposits in the Pilbara who had no viable way of exporting their iron ore.

For their part, BHP and Rio seemed content for the dispute to be tied up in litigation, in the hope that Forrest's grand vision would come to nought. The big miners argued that their rail networks were integral parts of a highly efficient mine, rail and port system, and that allowing others on the lines would reduce the amount of iron ore they could export. The legal battle dragged on for eight years, at a cost of tens of millions of dollars. It ended in a virtual stalemate, by which time Forrest had built his own rail lines. To this day, BHP and Rio have not had to carry one tonne of another company's ore on their Pilbara rail networks.

In 2003, Charles "Chip" Waterhouse Goodyear IV, the descendant of a wealthy American family that made its fortune out of tyres and railways, was appointed chief executive of BHP. Like Forrest, Goodyear could see the dizzying potential of China, which was on its way to becoming the world's biggest steel producer. But BHP, like Rio, had been too slow to expand its iron ore production capacity in the Pilbara and had been cutting back on its exploration work for several years. Goodyear set about boosting investment in new mines and by March 2004 had signed deals with Chinese steel mills that locked them into iron ore contracts worth $US9 billion over twenty-five years.

Goodyear, the boss of the world's biggest mining house, was in no mood to bother with irritants like Andrew Forrest who were talking up their chances of smashing the duopoly of BHP and Rio in the Pilbara. "You hear about new ideas every day," a dismissive Goodyear said. "Somebody threw a bag of rusty nails into their backyard ten years ago, they think they have an iron ore mine. It's gotten to the point where promoters are taking over this business and production's the secondary part."

Goodyear's "rusty nails" quote would inspire Forrest and his team to take the fight to the majors even more aggressively. Fortescue lodged plaints, or legal challenges, over BHP ground near the Chichester Range, claiming the miner had held the tenements for decades but failed to spend adequately on them. The move, although legal under the WA Mining Act, was generally frowned upon in the iron ore industry. But "plainting" was a method of obtaining mining leases and blocking rivals that Forrest had also used at Anaconda Nickel, which had lodged challenges over more than 100 tenements in the Goldfields.

BHP and Rio scoffed at Fortescue's prospects, telling journalists and analysts that the company's iron ore quality was poor and would never be accepted by the Chinese. Fortescue responded by arguing – correctly, as it turned out – that although much of its ore was of a lower grade, it had fast-burning qualities in sintering which would be welcomed by steel mills. The big miners also attempted to block Fortescue's groundbreaking plan to use special machines called surface miners, rather than traditional drill and blast methods. The surface miners had only ever been used on minerals that are up to ten times harder than iron ore. As it turned out, the German-made machines proved to be effective in cutting through the flat layers of rock at Cloudbreak. But Forrest, acting on the advice of mining boss Jim Williams, took a big gamble on their success: if the method hadn't worked, Fortescue would have faced a two-year delay and missed much of the iron ore price boom.

Forrest and his executives also came to believe that BHP and Rio were warning Australia's big investment banks against working for Fortescue or even publishing analysis of the stock. There is no direct evidence of this. However, in 2007, BHP dumped its long-standing broker, JP Morgan, after the firm acted as lead manager for Fortescue in a $504 million share placement. BHP and Rio did brief journalists regularly and convinced many of them that Forrest was doomed to fail. The former head of Rio's Pilbara iron ore operations, Greg Lilleyman, admits that BHP and Rio did talk down Fortescue, but he denies threats were made to investment banks. It's worth recalling, he notes, that most analysts will only cover a stock when a company has a tangible project, which Fortescue did not have until 2008.

Only one Australian mining analyst, the late John Veldhuizen at the small Sydney brokerage BBY, openly supported Fortescue in the early days. Forrest's old business partner, Warwick Grigor, never bought any Fortescue shares and was happy to tell journalists that he was suspicious of the speculative nature of the stock. "You can quote me as saying that when a stock rises on rhetoric and promotion, rather than hard-nosed facts and fundamentals, it guarantees a judgment day, eventually, that more often than not ends in tears," he said in 2005.

The mere perception that BHP, Rio and others were out to destroy them was enough to galvanise the Fortescue team and boost Forrest's efforts to create a culture defined by a David-versus-Goliath struggle. It wasn't a difficult analogy to make; Fortescue's market value in 2004 was a minuscule $60 million, whereas the combined worth of BHP and Rio was more than $100 billion. As John Clout recalled: "The more they threw at us the more we'd regroup, make sure we got our facts right and make sure that we got everything right and come back fighting, you know. It enormously energised the team. We were going to prove them wrong."

By refusing to share any of their infrastructure, BHP and Rio effectively forced Forrest to raise the money to build his own railways, which was to become the catalyst for much of Fortescue's future growth. Fortescue's company secretary, Mark Thomas, believes the big miners failed to understand the potential impact of their actions. "If they'd embraced us a little more, we wouldn't be what we are now," he said. "We'd be a little 20 million tonne per annum producer begging for space on their infrastructure. But by choosing to block us they forced us to become their nemesis."

# 8.

# BURKE'S BACKYARD

*I've met a lot of entrepreneurs,
but Forrest was the best I ever saw.*
—BRIAN BURKE

By 2004, Brian Burke had done his time in prison and was frantically reinventing himself as the most brutally effective lobbyist Perth had ever seen. The charismatic former Labor premier could open doors to ministerial offices and government departments at the drop of a hat – usually the well-worn Panama that would become famous a few years later when Burke was forced to explain his actions to a corruption inquiry.

Burke served two stints in jail during the 1990s – once for rorting his travel expenses while premier and again for stealing $122,585 in campaign donations (a conviction that was quashed on appeal). By the end of that decade he was desperate to make a quid. He went into business with his old Labor ministerial colleague Julian Grill, and the pair marketed themselves as Perth's lobbyists of last resort.

Some in Perth were convinced Burke had been unfairly treated in the wash-up of the 1980s WA Inc. scandals and that the charges brought against him had been petty. Everyone knew he was highly intelligent. As former prime minister Paul Keating said, Burke was "smarter than two-thirds of the WA Labor Party rolled together – that's why he keeps bobbing up". But plenty of others had never forgiven Burke for behaviour found by the WA Inc. royal commission to have been "improper" and "reprehensible".

The Labor Party's surprise victory at the 2001 state election was a godsend for Burke and Grill. Dozens of companies that needed help with government approvals were magnetically drawn to the pair. They boasted an enviable list of old mates and contacts extending from the outer reaches of the bureaucracy right up to ministers sitting around the cabinet table. Burke's influence within the Labor Party's factional system also meant he held sway over those members of parliament and ambitious staffers who needed help with fundraising or preselection. These networks of patronage meant he could easily call on favours as a lobbyist. It was a back-scratching exercise that took the art of political lobbying to new extremes. The memories of WA Inc. had been erased.

Grill, a Kalgoorlie boy, had excellent contacts in the mining industry, while Burke was a compulsive networker who was glued to his phone, cajoling ministers and public servants to get things done for his expanding client list. In 2006, his best year in business, "Burkie" racked up a staggering 25,389 telephone calls, faxes and text messages, according to the state's Corruption and Crime Commission (CCC), which was secretly bugging his phone lines for the entire time.

After being warned that Burke and Grill had begun to infiltrate his government, premier Geoff Gallop banned his ministers from contact with the duo in 2004. That simply forced them to adopt more subterranean methods of approaching decision-makers, although several Labor ministers did flout the ban. "Whatever the intention, we would relish another ban," Grill said in 2004. "The effect has been quite the reverse, as we have engaged a lot more clients since. Brian, in particular, saw it as a personal challenge to circumvent."

One of those clients was Andrew Forrest, whom Burke's magic would help to become a billionaire many times over. But when Burke first met Forrest, on 5 July 2004, the former premier was far from convinced about the young wheeler-dealer's chances of success with Fortescue. He knew the iron ore tenements the company had inherited from Allied Mining and Processing were far from impressive and he'd heard whispers in financial circles that Forrest was trying to make a quick buck rather than build a real company. But Grill, who had known Forrest from his time at Anaconda Nickel, had more faith in the mining promoter's abilities and convinced his business partner to meet Fortescue's executives to see if they could help.

Burke recalls that Forrest, at that initial meeting, was obsessed with gaining access to BHP Billiton's railways through the application he had recently made to the National Competition Council. But Burke says he told Forrest and his team that previous state governments, including his, had failed to force BHP and Rio Tinto to open up their infrastructure because it was too legally difficult. "I told him it's extremely complex. Sir Charles Court tried to do it and I tried to do it, but we both failed. It was clear to me that BHP was never going to let anyone on their railway line. I told him it was absurd to embark on a strategy of putting all your eggs in one basket, that he would be tied up in the courts for fifteen to twenty years. I said, 'You may win some rounds but BHP and Rio will not permit third-party access to their rail lines until every single legal challenge has been exhausted.' Well, you could have heard a penny drop. They all said to us: '"What should we do?"'

Burke told Forrest he needed what's known in Western Australia as a State Agreement Act – a special long term contract between the government and a company that allows mines and infrastructure to be built with government support. State agreements are usually difficult and time-consuming to negotiate because they send a strong signal to investors that a project has the government's imprimatur. The major iron ore mines established by BHP and Rio in the Pilbara all operate under such agreements.

Burke and Grill say they promised Forrest they could get Fortescue a state agreement to allow it to build its own railway, while providing the tiny company with a certain prestige that would boost its efforts to raise money and sign sales contracts. The pair signed on as Fortescue's lobbyists on a retainer of $10,000 a month – plus a success fee for negotiating the state agreement. They say they told Forrest they'd prefer to take the success fee in Fortescue shares, which were then trading at just 40 cents each. Forrest refused, offering $150,000 in cash instead. An angry Burke would never forget the millions of dollars he forewent as a result of Forrest's staunch refusal to hand over any shares.

Within a fortnight, Burke and Grill had made startling progress. On 22 July, the state development minister, Clive Brown, sent a letter to Forrest saying he would ask cabinet for approval to start negotiating a state agreement with Fortescue. When cabinet approval was gained, Burke and Grill went to work on the bureaucrats in various departments to sort out the details. But even the master lobbyists had trouble convincing some public servants that Fortescue had a genuine project that deserved the standing of a state agreement. "BHP had contacts in the department," recalls Burke. "They were saying, 'This company is never going to get up.' We kept running into obstacles that were BHP-inspired."

The Gallop government's decision to back Fortescue's bid to enter the iron ore industry had incensed BHP and Rio, who pointed out that they had invested billions of dollars in Western Australia over decades, employed tens of thousands of people and paid $550 million to the state in mining royalties that year alone. Rio believed that the approvals it needed to expand its Pilbara operations had been unfairly delayed because Department of

Industry and Resources staff had been redirected to work on Fortescue's state agreement. "Appropriate support should be given to companies that have invested billions of dollars in the state and have real projects and real plans, rather than promises," a senior industry source told the *West Australian* at the time.

Encountering resistance, Burke and Grill say they worked during 2004 to coax the bureaucrats into quickly drafting Fortescue's state agreement. "We got them to meet deadlines – we got a bit obsessive," Burke admits. By 28 October a draft agreement was in place and by 10 November the deal had been signed by premier Geoff Gallop. What normally took eighteen months had been done in less than four. The announcement of the deal helped to boost Fortescue's market value by a stunning $90 million in a single day.

But Burke and Grill – who worked closely with Fortescue's head of government affairs, Julian Tapp – still needed to get the agreement through parliament for it to have any legal effect. To achieve this, they faced a potentially major hurdle: parliament was due to sit for only a few more weeks before retiring for the summer break. Worse, a state election was to be held in late February, meaning it would be longer than usual before parliament would return, with no guarantee that Labor would retain power against Colin Barnett's Liberals.

Burke came up with a novel way of rapidly moving the Bill rapidly through parliament. He asked one of his old mates, Labor MP Norm Marlborough, to petition his colleagues for the Bill to be introduced in the upper house rather than the lower house, an unusual manoeuvre that enabled it to be passed in a single day. (Marlborough would later become a minister and kept a secret mobile phone purely for the purpose of communicating with Burke.)

When the Fortescue State Agreement Bill came before the lower house on the final day of the final week of parliament of 2004, Colin Barnett was outraged. The Opposition leader, who had served as resources minister in the government of Richard Court and prided himself on his knowledge of the industry, told parliament that such agreements were usually entered into only when the government was "absolutely confident" a project would proceed.

Fortescue, by contrast, had not even completed a feasibility study for its Pilbara iron ore project. An angry Barnett also complained that the Bill had been pushed through the upper house "a couple of hours ago" and he was now expected to vote on it without even a briefing from the government. The future premier was mystified as to why it was even being considered: "A state government agreement Act is a right to bank money, a right to raise money and a right to see a company's share market price rise. It is a signal to customers, investors, equipment suppliers and contractors that a project is about to happen. That is the history of agreement Acts. However, the history is that when an agreement Act passes this Parliament, it is a judgment by the government and the parliament that the project is at a sufficient state of development that it is about to proceed … It is not something thrown around as a political document in the run-up to an election. It is not something given to a proponent to wave at the sharemarket. The company can use its prospectus and media consultants to do that."

Barnett also raised questions about Forrest's troubled past at Anaconda. "I know that the proponent has an ability to raise funds," he said. "He did so on the Murrin Murrin project. I know that American bondholders lost $1 billion. I also know that the project got into serious difficulty. I know that a lot of people lost a lot of money on the sharemarket."

Norm Marlborough then leaped to his feet in defence of Fortescue's bid to take on BHP and Rio. "The leader of the Opposition needs to stop supporting all those foreign-owned companies in the Pilbara today that are running the iron ore industry,' Marlborough told parliament. "He ought to stop making excuses for the decisions that they make in their boardrooms in New York, London or wherever. Because the truth of the matter is that this [Fortescue] is a dinky-di Aussie company."

After some debate, Barnett said he would not oppose the Bill because he didn't want to jeopardise Fortescue's chances of developing the project. The legislation passed the upper house that evening – the fastest turnaround of a state agreement anyone had seen. It only emerged later that the government had negotiated the agreement purely on the strength of Fortescue's bullish media releases and the "binding contract" it had signed in August with China Railway Engineering Corporation to finance and build the Pilbara railway – a deal that was later revealed as merely a framework agreement without any details of price or schedules. The government had done virtually no due diligence of its own.

Burke cites the fast-tracking of Fortescue's state agreements – including a second one for the mine in 2005 – as a critical development in Fortescue's history. "All of a sudden it gave them huge credibility in China and in the US market," says Burke. Forrest became a big admirer of Burke's seemingly magical powers, referring to him in one meeting as a "national treasure". Burke remembers being surprised when Forrest took him aside one day to raise concerns about the lobbyist's weight, even offering him to make an appointment with his own doctor to help him drop a few kilograms. Says Grill: "Andrew got really close to us, and he particularly liked Brian."

Julian Tapp accepts that Burke and Grill did plenty of important work behind the scenes in negotiating the state agreement with public servants and that they had originally convinced Forrest to pursue the deal. But he says the lobbyists had less influence in the government than they claimed because many were still suspicious of them. "They had their own people they knew inside the government who they would deal with, but to be seen to be involved with them was actually likely to be slowing you down, not speeding you up," Tapp says. "I once made the mistake of taking Julian Grill with me to a meeting and I never did it again … because they engendered such hostility in some people."

Luckily for Burke and Grill, the state agreements were the first of many negotiations with the government that Fortescue would need to complete before it could start construction. Building an iron ore mine in the Pilbara in the mid 2000s was a vastly different proposition to the 1960s, when BHP and Rio had first charged into the region. Back then, there was no Environmental Protection Authority or green groups to challenge projects, no native title laws to worry about and no need to pay consultants to carry out Aboriginal heritage approvals or liaise with local communities about a project's "social impact".

In early 2006, former ABC journalist Alan Carpenter became premier, replacing Gallop, who was suffering depression. Carpenter immediately dropped the ban on ministerial contact with Burke and Grill, but it was a decision he would come to regret deeply. In a further stroke of luck for the lobbyists, Carpenter appointed another former ABC reporter, John Bowler, to the influential role of resources minister. Bowler was close to Grill, who had preceded him as the MP for the seat of Eyre in the Goldfields. He also got on well with Burke, who had donated $2500 to Bowler's re-election campaign in 2005. Bowler's workload in 2006 was heavy; the great mining boom was in full flight and every company with the sniff of a viable project was spruiking its plans and trying to win government approval.

Each week, Bowler experienced what he referred to as the inevitable "FMG Friday crisis", involving a problem that Forrest needed addressed by Monday. If it wasn't fixed, Bowler conceded, there would be "hell to pay". When he appeared before the CCC to answer questions about his close links to Burke and Grill, Bowler made no apologies for cutting a few corners in pushing through Fortescue's approvals. "This was a multibillion-dollar project that has – under my guidance and help, along with other ministers – been driven quicker than anyone could imagine," he said. "Andrew Forrest drove it, pushed everyone, pushed his staff, and sometimes we helped him along as quick as we could. I don't see anything wrong with that. That is my job as minister, to help people, and if I can cut a corner – maybe it's my attitude to do so, that's my natural instinct, and it might get me into trouble sometimes."

Bowler cut as many corners as he could for Forrest, but by the middle of 2006 there was still one crucial approval that Fortescue needed to get its project started. The company wanted to build its railway line to Port Hedland through a 2500-square-kilometre patch of land that was shielded from development under the Aboriginal Heritage Act. This was a potentially major obstacle for Fortescue because the region, known as Woodstock Abydos, had been given the highest level of protection available under state law. It contained some of the oldest and richest indigenous rock engravings in the world, as well as many Aboriginal ceremonial sites, artefacts, stone arrangements and rock shelters.

Western Australia has a woeful record in protecting the internationally renowned Aboriginal rock engravings of the Pilbara. In the 1980s, Woodside Petroleum was allowed to destroy or relocate thousands of rocks on the Burrup Peninsula to make way for its huge North West Shelf gas project. To this day, many of the 30,000-year-old pieces of rock art that were cleared remain stacked behind a cyclone fence. The artefacts dumped there are thought to be twice as old as the Pyramids or Stonehenge. The rocks that were not removed are subject to deterioration, theft and vandalism, leading the New York-based World Monuments Fund to declare the Burrup rock art one of the 100 most endangered heritage sites in the world.

At Woodstock Abydos, which lies about 200 kilometres inland from the Burrup Peninsula, BHP had already been granted special approval to build its railway to Port Hedland. This and other approvals prompted the federal environment department to sound the alarm in 2011 over the intrusion of industry into the area. "The Woodstock Abydos experience is perhaps one of the most striking examples of development incrementally disturbing an area of recognised outstanding heritage significance," the department found.

> There are many sites very close to rail tracks and maintenance roads, so dust accumulation on rock art poses an ongoing, serious threat. Sites suffer from neglect, poor fencing and lack of protective measures. There is no program of monitoring of the sites or individual images, and there have been reports that additional rail corridors are planned in the years ahead. Woodstock Abydos shows that even the highest form of protection available for Aboriginal heritage sites under Western Australian law may not be a guarantee of protection, and that individual approvals can have a serious cumulative adverse effect.

In early 2006, Fortescue asked the indigenous affairs minister, Sheila McHale, to excise a corridor of land from the Woodstock Abydos protected area measuring about 200 metres wide and fifty-six kilometres long. But McHale – who was not as helpful as her cabinet colleague John Bowler – sat on the application for months without making a decision.

Brian Burke phoned Bowler, the resources minister, and asked him to tell McHale that her indecision was holding up Fortescue's entire project. But McHale said she was planning to reject the application because, as Bowler reported back to Grill, she believed allowing the railway would set an "unhealthy precedent". As McHale later told the CCC, she also felt uncomfortable that Bowler was pushing her to make a decision at the behest of Burke and Grill. She'd probably have felt even more uneasy if she'd known Bowler was relaying most of their confidential discussions straight back to the lobbyists.

McHale referred Fortescue's application to the state's Aboriginal Cultural Material Committee, an advisory body that evaluates the importance of indigenous sites and makes recommendations to the government. On 7 June, the committee recommended to McHale that she should reject Fortescue's application. This was a huge blow for Forrest. Making matters worse, a group of Aboriginal custodians, led by renowned author Sally Morgan, had been made aware of Fortescue's plan to build through Woodstock Abydos and had been campaigning against it.

Fortescue decided to crank up the pressure on the government to ignore the ACMC's recommendation. Julian Tapp, as head of government affairs, sent Bowler an email on 16 June saying the company needed a decision urgently because it was running the risk of being unable to obtain financing for the project. "The project will be killed by indecision – we need a decision now," Tapp wrote.

Nobody seems to have queried why Fortescue could not have simply spent a few extra million dollars to divert its railway line around the protected area. Forrest would later claim this to have been virtually impossible. "We'd have had to go around [Woodstock Abydos], which would have been another set of applications, put the project back two years and we'd have gone broke," he said.

Faced with this apparent crisis, Forrest decided to take matters into his own hands. In a meeting with McHale on 7 July, he used all his persuasive powers to try to convince her to allow the railway. Forrest promised McHale that Fortescue would use non-explosive blasting techniques in its excavation work that would not damage the ancient engravings nearby. "I said, 'Minister, each time they go to blast, I will personally stand in front of the carvings so if I happen to be wrong, I will get hit and not the carving. Do we have a deal?' And she said, 'If you can commit to that, we have a deal.'"

Significantly, McHale appears to have agreed to meet Forrest only after the intervention of the premier, Alan Carpenter. Evidence to the CCC showed Bowler had personally raised the issue of Fortescue's delay with Carpenter. Later, in a secretly recorded phone conversation, Grill said Bowler had been responsible "for getting Alan to put a bit of pressure on Sheila to come up with the right sort of answer".

It's not known whether Forrest ever fulfilled his promise to act as a human shield for the rock art during the railway construction. But the pledge must surely have helped his argument. Three days later, McHale approved Fortescue's application. The government's respected planning minister, Alannah MacTiernan, then signed off on the final route for the railway. When Fortescue's first train eventually ran down the track in 2008, it was called the "Alannah MacTiernan Express" and Forrest stood on the cowcatcher waving the Australian flag.

On 10 July, the same day as McHale's critical decision, Grill rang Forrest to discuss their successful lobbying efforts. The conversation was being covertly recorded by CCC investigators. The recording shows Forrest was concerned Grill should not "leak" McHale's yet-to-be-announced decision and that he viewed himself as a person who "loves" Aboriginal rock art.

FORREST: Hello.
GRILL: Oh, Andrew, Julian Grill. How are you?

FORREST: G'day, mate, how are you?

GRILL: Yeah, good, I heard from inside the Department of Indigenous Affairs that Sheila's virtually reached agreement with you in respect to the basic elements of the rail route on Friday. I just wanted to find out whether that was correct, as to whether we needed to take any further action, just what the score was.

FORREST: No, look I hope not, Julian, hope that we can just go very quietly now, let the government announce it when they're ready. We really got to make sure, Julian, it does not leak, 'cause Sheila's petrified about that. What she did is put us through a very, very fine-mesh sieve, which ... suits me fine as a West Australian who loves rock art, we've been able to guarantee and enshrine the safety and protection of every single carving ... and still not hassle the railway line particularly.

GRILL: Oh, that's good. It's just that it seemed to me from the description I was getting that you might have to deviate it somewhat but ...

FORREST: No. What was given away was large chunks of the 200-metre corridor. And I'm comfortable with that.

GRILL: Alright, then, so we'll leave it on that basis, then?

FORREST: Yeah, I think so. Just make sure, mate, no one leaks the fucking thing, please.

GRILL: Yeah, sure. And you know, as far as I'm aware, the only person that's been informed is myself. I am not aware of anybody else. Of course there'll be other people within government, you know that, don't you?

FORREST: Other people within government but Sheila is ... I think it is a very brave decision by her. She's taken a very long time to reach it and strung this out to a point where, you know, I had decided the project will break. It can't stretch ...

you know projects, they usually just stretch and stretch. We will actually go under ... and she considered it all.
GRILL: Oh, you're very persuasive too.
FORREST: Well, the facts really helped.

As a result of its investigation, the CCC found Bowler had taken advantage of his public office by passing on confidential information to Burke and Grill that he gained from his discussions with McHale and Carpenter. But it also found there was no evidence that Bowler's actions constituted misconduct. By the time of the CCC's finding on this issue in 2009, however, Bowler had long resigned as a minister over his close links to the lobbyists and Forrest had greatly benefited from the Burke–Grill connection by winning all his approvals and building the whole project.

As soon as the CCC hearings into Burke and Grill began in late 2006, Forrest ran as fast as he could from the pair. He told journalists that Fortescue had engaged Burke and Grill only to train Julian Tapp, who was unfamiliar at the time with the state bureaucracy. "As far as any lobbying goes, they don't do that for us," Forrest said. "When we need to speak to a minister or a member of the bureaucracy, we do that." He repeated this claim to the *Australian Financial Review* a few months later. "We did retain Burke and Grill to show our new government relations manager the ropes, and they did that effectively. But we did not ever use them in our approaches to government," he said.

Forrest was either mistaken in making these statements, or else he was deliberately telling untruths. Perhaps his comments showed how desperate he was to distance himself from the stench of corruption that had suddenly enveloped Burke, Grill and their mates in the state government. As evidence given at the CCC would show, Burke and Grill did have extensive contact with John Bowler and senior public servants on behalf of Fortescue. Grill also now confirms he had direct contact with McHale in relation to the Fortescue rail application. Burke and Grill had also lobbied Labor MPs and senior bureaucrats to push Fortescue's state agreements through parliament during 2004 and 2005 – a point Forrest must have temporarily forgotten when journalists quizzed him about how well he knew the pair.

As soon as the media storm erupted over Burke and Grill in 2007, Forrest ceased all contact with them. These days, the pair are still bitter at the way their contract was ripped up. "Most clients took the time to express gratitude for the work we'd done, and explained they could no longer use us, which was fair enough in the circumstances," Grill says. "Forrest just cut us off." Grill's wife, Lesley, sent bills to Fortescue seeking payment for the $10,000 owing for the previous month's work. She followed up with emails but never received a reply.

Within weeks, Brian Burke's name had also infected federal politics as John Howard's Liberal government attempted to drag the Opposition leader, Kevin Rudd, into the controversy. Rudd had been the VIP guest at a Perth dinner Burke had organised in 2005 to introduce him, as Labor's foreign affairs spokesman, to Burke's best clients. Forrest sat alongside Rudd at the dinner, which was held in the private dining room at Perugino restaurant in West Perth. At the height of the Burke paranoia, federal treasurer Peter Costello unwittingly insulted Forrest, and plenty of other Perth business leaders, by gleefully telling parliament: "Anyone who deals with Brian Burke is morally and politically compromised!" But Costello's own party was soon dragged into the mess when WA Liberal senator Ian Campbell, the federal minister for human services, was forced to resign after admitting he once chatted to Burke for twenty minutes.

Fast-forward six years and Brian Burke is reclining in a chair in Julian Grill's plush apartment, which overlooks the city they once dominated. The pair have agreed to discuss their work for Forrest – one of their best clients – but Burke seems especially keen to rehabilitate his name. He says his lobbying business has been dead ever since the CCC began examining his lobbying efforts for Fortescue and scores of other Perth companies. The investigation, he claims, was purely a witch-hunt that was designed to destroy him, but that only produced a string of failed prosecutions.

Like others, Burke believes BHP's failed attempt to cut Forrest down became the catalyst for his success. "BHP made a fundamental mistake by wrongly assessing the character of Andrew Forrest," he says. "BHP did everything they could to stop him; they brought every bit of their firepower to bear. They basically created him. If they had gone easy on him, he wouldn't have fought back so hard."

Reflecting on his work with Forrest, Burke says he soon overcame his initial scepticism and came to realise that Forrest was a freak – even better than the deal-makers he knew well in the 1980s, including Alan Bond and Robert Holmes à Court. "I don't think anyone else but Twiggy Forrest could have gotten that company off the ground," says Burke. "I've met a lot of entrepreneurs but Forrest was the best I ever saw. He was the best salesperson, the best driver, the best demander. He is the best at exuberance I've ever seen. He got people to do things they thought they could never do. He would treat people in ways that were not the best, but maybe that's an ingredient of his success. He's unique – but that doesn't make him likeable."

# 9.

## TO GET RICH IS GLORIOUS

*You will look back and know you were part of history,
like building the Sydney Harbour Bridge.*
—ANDREW FORREST

Although he was a superb salesman and entrepreneur, Forrest could never have realised his dream alone. He badly needed capable executives around him and, like any boss, he relied on a mix of luck and skill in finding them. In Eamon Hannon, David Mendelawitz and John Clout, Forrest had found people who understood iron ore and geology better than most. Most of his other early appointments were also successful, although some had to be convinced to leave safe jobs to join Fortescue.

One day late in 2003, however, a young man Forrest would soon look upon as crucial to Fortescue's success actually came looking for him. David Liu had read about Forrest's idea to start a mine and phoned Fortescue's office in Perth looking for a job. Liu, who had recently finished university in Perth, was descended from a long line of steel mill executives in Nanjing and was keen to join the iron ore industry in Australia. During the phone call, he discovered that Forrest and his senior executives were at a steel conference in the Chinese city of Maanshan. He flew straight there and began helping Forrest voluntarily, in the hope that he'd be put on the payroll back in Perth. It was the same method Forrest had used to break into stockbroking in the 1980s. Liu was soon assigned to Fortescue's iron ore marketing team – his first real job apart from teaching English.

Forrest asked Liu and others to fly to China and demand pre-payment from the steel mills for iron ore that hadn't even been pulled out of the ground. In the world of iron ore sales, this had never before been attempted – and at first it proved impossible. Philip Kirchlechner, then head of marketing, recalls: "Working for a junior, it's terrible – people don't want to talk to you and don't take you seriously. And there was all of Forrest's baggage from Anaconda, so it was an uphill battle." Executive director Russell Scrimshaw, who made more than 100 trips to China in only a few years while overseeing Fortescue's commercial operations, says it was imperative that the company secure long-term sales contracts if it was to have any chance of raising the money to build a mine, port and railway.

While Kirchlechner used his high-level contacts to market iron ore to the larger steel mills in China and Japan, Liu went to work on selling the Fortescue vision to the small end of the industry. Liu and Scrimshaw adopted Forrest's foot-in-the-door approach to business, often sitting in the lobby of a Chinese steel mill until someone would come and see them. But in the early days, few were interested in talking to them. On one embarrassing occasion, Scrimshaw recalls, the only person they managed to speak to about their iron ore was the cleaner: "Every time we would go to this mill, we thought we had meetings but we'd get there and the key guy had gone off to a conference or gone interstate or suddenly disappeared. So David and I just decided this was ridiculous, we decided we were going to sit in the lobby of their head office until someone of note came to see us. Unfortunately, it was late afternoon and we were getting bored with this little strategy that didn't seem to be working, and this person came in. We were engaged in a good discussion and I thought, 'We're getting somewhere here,' but it turned out he was the cleaning supervisor and he was just moving in to start work for the night. So David and I went from this moment of, 'We're making some headway here,' to, 'This is ridiculous.'"

As time went on, Fortescue raised its profile in China and the marketing task became easier. As it turned out, Fortescue's timing was ideal because the Chinese mills were becoming desperate for a new iron ore supplier to break the stranglehold of BHP Billiton, Rio Tinto and Brazil's Vale. "You had three players supplying the majority of the world's iron ore and there were a couple of hundred steel mills, so the odds were weighted heavily in favour of the supplier," Scrimshaw says.

The smaller Chinese mills were the first to sign up. At the time, they were being forced to pay premium prices for iron ore through the high-priced cash market because the traded ore market was dominated by long-term deals between the three big suppliers and the major steel mills. Scrimshaw recalls: "When someone else came along and said, 'Okay, I'll supply you at the benchmark price,' the smaller guys said, 'Where do we sign?'"

Forrest became a devotee of Liu, who handled many of the big negotiations with China. Like Forrest, Liu was a born hustler who knew how to tell a good story. "David Liu is the guy who … personifies courage and would go into these steel mills or ring these people up who wouldn't take a call normally from a little twerp, which is how they have seen him," Forrest said. "But he'd ring them up with authority in his voice representing Fortescue and speak with such courage about what we were going to achieve, when it was just a game plan."

Liu and his colleagues had one big advantage: by 2005, Chinese demand for iron ore was swelling at a rate that neither Forrest nor anybody else had anticipated. The Chinese steel industry was also angry with BHP, Rio and Vale over what it saw as their bullying tactics in rapidly forcing up the price of iron ore. Forrest skilfully cast himself as the good cop of iron ore mining, as opposed to the greedy triumvirate which controlled more than 75 per cent of global exports.

Efforts by BHP and Rio to talk down Fortescue were backfiring badly on them. When they told the heads of China's biggest steel mill, Baosteel, not to believe anything that Forrest uttered about his new project, the executives demanded to know more about the new entrant. In 2004, Baosteel signed a memorandum of understanding to buy iron ore from Fortescue, an agreement that was later converted into a long-term offtake contract for 5 million tonnes a year. By the middle of 2006, more than twenty sales agreements had been signed for almost all of Fortescue's planned annual output of 45 million tonnes of iron ore.

All the offtake deals in the world, however, were worthless unless Fortescue could find the big money needed to build the project. This is where Forrest's promotional skills were fully tested. When cash was again running low in 2005, he convinced billionaire New York hedge fund boss Phil Falcone to invest tens of millions of dollars. Falcone later built up a 16 per cent stake in Fortescue that was worth billions of dollars at the height of the mining boom. Forrest regarded Falcone, who owned Harbinger Capital, as one of his most loyal supporters – he had also invested in Anaconda – and he saw plenty in the money manager's risk-taking approach that he admired.

But like so many others Forrest had befriended, Falcone's reputation was about to take a turn for the worse when he lost billions of dollars in a series of bad investments. The US Securities and Exchange Commission later laid fraud charges against Falcone, alleging he used $113 million of his clients' funds to pay his personal taxes, and that he manipulated bond prices and violated trading rules intended to stop manipulative short selling. His fall from grace was complete in May 2013, when he agreed to a two-year ban as an investment adviser. By then, Forrest had nothing to do with Falcone, who had dumped his entire stake in Fortescue.

After securing Falcone's support in 2005, Forrest brought in another American investor, the secretive hedge fund Leucadia International Corporation. Leucadia kicked in $US400 million for a 9.9 per cent equity stake in Fortescue and lent the company a lifesaving $US100 million. The dazzlingly high interest rate on the Leucadia thirteen-year convertible note – equating to 4 per cent of Fortescue's future sales revenue – was a reflection of Forrest's short-term need for cash and his desperation to attract support from financiers. Before throwing their money at Fortescue, Leucadia's shrewd founders, Ian Cumming and Joseph Steinberg, had asked hard questions about Forrest, and they liked what they saw. In a letter to shareholders the following year, they explained their investment: "FMG is the creation of a hyperactive, smart, energetic Australian entrepreneur named Andrew Forrest; imagine the Energizer Bunny," they wrote.

Forrest still needed to raise another $US2 billion to fund construction of the project. But every financing scheme he tried had failed. A deal for Chinese construction companies to build the mine, railway and port had collapsed spectacularly in March 2005, sparking an investigation into Forrest's conduct by the Australian Securities and Investments Commission that led all the way to the High Court (see Chapter 11). Forrest also held funding talks with Indian steel magnate Lakshmi Mittal, but pulled out of a proposed deal because he feared the tycoon would seek control over Fortescue, just as Anglo American had done at Anaconda. The big Australian banks were not an option for Fortescue because none wanted anything to do with Forrest – they were suspicious of his business methods and were still scarred by their experiences with Perth's brat pack of deal-makers in the 1980s.

The last hope for Twiggy – as it had also been at Anaconda in 1997 – lay with the high-yield bond market. Only those with a taste for risk would back him. A prominent Jewish banker in New York, Eddie Sugar, whom Forrest had known from his time at Anaconda, helped promote the Fortescue story to US investors, who were more prepared than their Australian counterparts to back an entrepreneur who had previously failed.

Over three hectic weeks in the middle of 2006, two teams of Fortescue executives travelled to fourteen cities across the world. The "blue team" of Forrest, finance executive Peter Thomas and John Clout headed to Europe, while the "red team" of Scrimshaw, Graeme Rowley and Chris Catlow went to the United States. In the middle of the trip, they crossed over in Hong Kong. During a blur of endless breakfasts, lunches, dinners and boardroom presentations, the Fortescue executives pitched their plans to hundreds of fund managers. They were accompanied by executives from global bank Citigroup, which, by accepting the role of lead manager to the bond issue, had become the first mainstream institution to back the fledgling company.

The fundraising roadshow encountered several near-fatal setbacks. A week before the bond issue was due to be finalised, Standard & Poor's issued a credit rating for the notes that was two levels below what it had previously indicated, a move that effectively killed off the financing. Scrimshaw recalls he found out about the Standard & Poor's rating late one night in New York. "I went to bed at 1am thinking, 'We're finished here, we won't be able to raise the money.'"

Forrest didn't go to bed that night. Instead, he stayed up and convinced the Standard & Poor's executives in New York, the United Kingdom and Australia of the untold potential of his project. They only relented after several hours of his ear-bashing, agreeing to boost the rating on the bonds by two levels to BB.

That was still three levels below investment grade, but it was enough to attract those risk-taking funds attracted by the high yield. Scrimshaw recalls that the Citigroup bankers in New York were amazed by what Forrest had achieved. They had seen bonds talked up by one level before, but never by two.

Just as the raising appeared to be getting back on track, however, liquid explosives were found at London's Heathrow Airport and Israel invaded Lebanon in retaliation for the kidnapping of two soldiers. These events spooked global markets, leading to one major investor threatening to pull out of the bond deal unless key changes to the terms were made. With only fifty hours remaining until settlement, Citigroup and the bondholders reached an agreement.

But even that wasn't the end of the drama. With a few hours left, the bond raising looked likely to fall about $150 million short. Forrest needed to move fast. He called an old contact, Richard Handler, the head of US investment bank Jefferies, to ask if he would underwrite the balance. Recalled Scrimshaw: "There was a silence that seemed like forever, with our hearts beating hard because we knew we were playing our last card. And he came back after what I thought was forever, it was probably ten seconds, and said, 'Andrew, I'd be happy to help you.' So we looked at the Citigroup guys, they looked at us, and within minutes we were watching the bonds be posted."

Aided by the blood, sweat and tears of his troops, Forrest had raised $US2 billion, ensuring that his reputation as a master promoter and fundraiser remained intact. It was the biggest high-yield bond issue ever to come out of Australia and one of the biggest bond raisings ever seen in the mining industry. What's more, most of the funding had come from American investors, suggesting the US bond market had forgiven Twiggy for his Anaconda sins.

One of Perth's top investment bankers, Eddie Rigg, said after the record-breaking raising that he believed Forrest would become known as the "greatest visionary in the Australian resources sector". The secret of his money-raising genius, Rigg suggested, was an ability to pitch a story on such a huge scale to attract the major-league financiers, which nobody else had the "balls" to even contemplate.

Rigg knows from personal experience that behind Forrest's bright façade is a ruthless businessman prepared to do almost anything to realise his ambitions. A year before the 2006 bond raising, Forrest sued one of Rigg's clients over the rights to a massive iron ore deposit in the Pilbara. The Federal Court heard evidence of a crucial phone conversation between Rigg and Forrest, the details of which were heavily disputed. But the judge threw out Fortescue's case. Rigg now says of Forrest: "How he achieves things is questionable, but are the people of Western Australia better off because of his vision? Absolutely."

Another key ingredient of Forrest's success lay in his impeccable timing. Fortescue had raised the money in 2006 as the global iron ore price was soaring and international investors finally began to accept that China's seemingly inexhaustible hunger for commodities was part of a "resources super cycle". The predictions Forrest had made about China's awakening at the Celtic Club in West Perth had been realised in just three years.

As the money began pouring into Fortescue's bank accounts, work began on building a mine capable of producing 45 million tonnes a year of iron ore, a 256-kilometre railway and a deepwater port. This sounded like a big undertaking, but for Forrest it was the easiest part of the challenge. He knew that an iron ore project was far easier to develop than the technical complexities he had encountered at Anaconda.

In essence, iron ore mining is a massive earthmoving operation supported by a logistics business. Yet it is hardly effortless. Large trucks must first remove a twenty-five-metre layer of unwanted dirt, known as overburden, that sits above the prized ore. Fortescue's surface miners then cut up the ore and feed it into waiting trucks. From there it is crushed and the impurities removed, before it is conveyed to a train loader, which piles the dirt onto trains. After the rail journey to Port Hedland, the ore is stockpiled until it is picked up in huge scoops and taken to the wharf on conveyors. A ship-loading machine then pours the ore onto massive vessels, which set sail at regular intervals for the hungry steel mills of China.

The biggest problem for those tasked with building the infrastructure was that they had only a $2-billion budget when it appeared they would need at least twice that much. By 2006, the cost of labour, equipment and just about everything else in the Pilbara was skyrocketing as the resources industry began to crank up new projects in response to Chinese demand.

Forrest set up a special team, led by Alan Watling and Peter Thomas, to find new ways of doing things that would slash the cost of the project. After four months of intense meetings, the team delivered a project plan with a budget of less than $2 billion. Peter Meurs, as the head of main contractor Worley Parsons, was taken aback at Forrest's ability to get people to believe in improbable targets. "All of the project management things we'd learned, all of the experience [was] thrown out the window," he said.

Forrest's answer to the skills shortages that were engulfing the industry was to attempt to differentiate Fortescue from other miners competing for the same limited pool of labour. He began speaking of his "Fortescue family", a leitmotif that helped create an almost evangelical aura around Forrest and the company. Fortescue's open-plan office on Perth's Adelaide Terrace also fostered an air of informality and egalitarianism – values Twiggy had long seen in himself. Forrest's own cubicle, right in the middle of the office, was exactly the same size as that of the junior accounts clerk. The Fortescue office was deliberately on the eastern outskirts of the CBD, at a safe distance from the towers of BHP, Rio and the rest of the establishment.

With characteristic flourish, Forrest gave an insight into the unique culture he was trying to create while addressing a group of employees in the Pilbara: "Everyone here is part of our Fortescue family. You are family as soon as you step foot on to one of our projects. And what families do is love one another. Now I know that might sound a bit odd to some of you blokes. But that's what you do at Fortescue. You love one another, you look after your mates, you care for one another. And if you are having personal problems, any problems, you tell someone. Don't bottle it in. If you have an idea how to do it better, you tell someone. Or you email me. That's the Fortescue culture. You will look back and know you were part of history – like building the Sydney Harbour Bridge."

Those who worked closely with Forrest found his energy and penchant for risk intoxicating. They were pushed to achieve things most never believed they could. Employees in their twenties were handed responsibilities for overseeing key parts of the project they would never have been given at other mining companies. But many also found Forrest infuriating. He had a scattergun approach, which meant he was constantly coming up with big ideas that needed to be acted upon, only for them to be ditched soon after. He also had a short attention span. Says a former colleague: "Because he's a man of action, he doesn't listen too often – he wants to know what the problem is, what the proposed action is, and if he doesn't agree with you he'll tell you what he wants to happen. He doesn't suffer long descriptions and presentations and long documents. He wants things distilled."

Many colleagues also sighed at his propensity to stretch the truth and to take credit for achievements when others deserved the recognition. Forrest would also habitually keep his counsel. Says the former executive: "One thing I found annoying is that he always knew more than you did. It's a style that enables him to keep control of what he's doing. He often knows little bits of the picture and you might have another part of it, and he'll deliberately hold back those bits to his own advantage. I used to find that incredibly frustrating."

Some executives fell out badly with Forrest. Tensions developed with marketing boss Philip Kirchlechner over his reluctance to fall into line with the gung-ho style of the company. "If I was a little bit hesitant about telling the Fortescue story, or not optimistic enough, Andrew would say, 'Philip, that's negative marketing,'" he recalls. "And if he really wanted to insult me, he would call me conservative."

Kirchlechner left the company in May 2006 because, he claims, Forrest reneged on a verbal promise to give him 300,000 shares a year. The pair had a shouting match in the office before Forrest ordered him to leave without taking anything from his desk apart from his coat. By the time Kirchlechner was standing on the pavement outside, his BlackBerry had already been deactivated. It was a warning to others that Forrest could never be beaten in an argument. "He always thought he was right," Kirchlechner recalls. "I had so many arguments with him and he would say, 'I know I'm right.' He had more conviction about himself than anyone else I've ever met."

Other executives were also bitter at not having shares or options despite their years of loyalty to the boss, whose own fortune was growing at high speed. Ed Heyting, a senior manager charged with building infrastructure, left in 2005 after a dispute over options. The hard-nosed Forrest also got rid of anyone he regarded as a naysayer. "Their desk would be suddenly empty and they would just evaporate," says a former executive.

Forrest admitted he couldn't tolerate "cynics" working under him. "That will bring you undone," he told a business audience in

2006. "Anyone who is not part of that unified vision should leave. It's not all soft and cuddly. I learned the importance of team building at an early age while mustering and found out pretty quick that unless everyone was on the same page, it only needed one who wasn't to bring the whole show undone." Despite his ruthlessness, some employees also recall Forrest's genuine acts of kindness, including visiting injured or sick workers in hospital and offering to help colleagues move house.

Forrest also wouldn't tolerate being told that something was impossible, even when engineers who knew much more than him were saying it couldn't be done. Alan Watling, who had been with Fortescue since the beginning, finally snapped one day in 2007 when Forrest demanded that he build the railway several months faster than previously agreed. Watling, one of the few people to regularly stand up to Forrest, began yelling at his boss in the middle of the office: "Remind me, Andrew, how many fucking railway lines have you built?"

By the middle of 2006, Forrest had become obsessed with building the project in record time, both to meet the sweet spot of Chinese iron ore demand and to avoid running out of money. He installed a big timepiece in the foyer of Fortescue's Perth headquarters. In glowing red digits, the clock ticked down the days, hours, minutes and seconds until the first batch of iron ore would leave Port Hedland. The target date had to be extended by a few months when Cyclone George tore though the Pilbara in March 2007, killing two workers and destroying Fortescue's rail construction camp at a cost of $100 million.

The countdown clock in the office unnerved many employees, but it also drove many of them to work harder. Forrest tried other tactics too. For the two years leading up to the maiden shipment in May 2008, he ordered in big plates of sandwiches at lunchtime so workers wouldn't need to leave their desks. As the deadline loomed ever closer, he even began discussing with colleagues the need to build a crèche in the office to allow employees with children to be in the office for longer hours.

The stress of working with Forrest took its toll on many people. Some left because they could not handle the relentless pressure and demands to put in long days. Forrest would send emails to executives in the middle of the night and expect a response. Most employees struggled along without ever taking a holiday, convinced they were part of something special. David Mendelawitz, who became Fortescue's head of business improvement, recalls he and his wife had a week's holiday in Thailand after he'd been working for several years without a break. He could do virtually nothing but sleep for the whole week. But he believes the only way Fortescue could have been created was through the superhuman efforts led by Forrest. "Everyone told us we would never succeed; the guys that survived were the ones who said, 'We will make it succeed,'" he says.

Forrest's goading of his staff paid off in the end. Within two years the whole project had been built – a shorter period than it might have taken to build a large house in Perth at the time. Forrest knew the deadline had to be met because Fortescue would have run out of money if had been unable to start selling iron ore. In the end, the total bill for the project was $2.8 billion – higher than the earlier budget estimate but still good value, given the cost blowouts that were rife in the industry at the time.

By May 2008, Fortescue had a new port at Port Hedland with two berths catering for capesize vessels – the huge ships that are too large to pass through the Panama and Suez canals. And it had a brand new railway, consisting of 420,000 sleepers and 38,000 tonnes of steel, which ran right next to BHP's tracks for most of its length. For the first time in forty years, a new company had built a rail line in the Pilbara. Naturally, it was done with record-breaking haste; the rail construction crew claimed an Australian record when it laid 3.3 kilometres of track in a single day in March 2008. Fortescue bought more than 800 ore cars made in China, and a fleet of fifteen locomotives arrived from the United States. When the first iron ore at Cloudbreak was loaded on the trains, the project was ready to go. In total, there had been 10 million work hours, 4500 procurement orders, 13,000 engineering drawings, 330 contracts and 24,000 invoices issued to get the mining project started.

On 15 May, the clock in Fortescue's Perth office ticked over to zero. On a clear day in Port Hedland, Forrest, surrounded by his family and some of his closest friends, including fellow billionaires Kerry Stokes and James Packer, stood on the wharf as the carrier *Heng Shan* began loading the first 180,000 tonnes of iron ore to be shipped to China. For executives like Russell Scrimshaw, this was the moment that made all the turmoil of the previous five years worthwhile. "Nothing short of a miracle," he said of the moment he watched ore being loaded onto the ship. "There isn't another example anywhere in the world of a resources company starting from absolutely nothing, one person with an idea, no money, no iron ore [and] building something so quickly."

When the *Heng Shan* arrived in Shanghai after a ten-day journey from the Pilbara, Forrest was visibly emotional, hugging strangers and toasting the event with glasses of Margaret River red wine. He acknowledged that he had been in the right place at the right time. "It's the end of five years' hard work which could really have only happened in these five years," he said. "It needed a confluence of historical events which Fortescue was lucky enough to capture."

Well before that first shipment, however, Forrest's world had changed significantly. As investors began to recognise that the one-time cowboy had to be taken seriously, Twiggy became the poster boy of the Australian mining boom. Forrest's self-belief

John Forrest's 1869 expedition (left to right): Malcolm Hamersley, Tommy Windich and John Forrest, who was only twenty-one. Although the expedition failed to discover land for agricultural or pastoral use, Forrest's compass was affected by the presence of minerals in what would become known throughout the world as the Goldfields. (State Library of Western Australia.)

Lady Forrest and Sir John Forrest in 1909, when he was federal treasurer. Forrest's greatest disappointment in politics came when he failed to win the leadership of the Liberal Party – and most probably the prime ministership – by a single vote in 1913. (State Library of Western Australia.)

Coach: G.L. Matthews
Back: A.J.D. Smith, R.C. McKay, J.F. Douglas-Smith, A.D. Golding, K.C.M. Audas, W.H. Breidahl, W. Carrington-Jones, P.J. Morris, P.M.J. Hanlon, A.H. Dixon, A.J. Keay, M.I. Cearns, D.R. McComb.
Centre: M.P. Herman, J.I. Barrett-Lennard, D.G. Agnew, P.W.R. Lipscombe, G.M. Hall, R.A.S. Infirri, A.J.H. Forrest, B.D. Carrington, P.J. Kerr, M.R. Alderson, J.R. Hawson.
Front: S. Rigney, T.H. Lemann, J.G. Evans, C.B. Oliver, R.P. Cullingford, G.M. Collett, G.I. Bowman, J.L. Allwood, G.W.H. Crommelin, M.J.R. McTavish, R.E. Pixley.

Andrew Forrest as a member of Christ Church's 1973 prep school swimming team. He is in the middle row, fifth from the right. Forrest struggled at Christ Church and ended up finishing his secondary education at Hale School.

Forrest (back row, far right) became a prefect in his final year at Hale School in 1979. Also pictured are his close friends John Morrison (front row, far left) and Simon Lill (back row, fourth from left). Forrest credits headmaster Ken Tregonning (front row, centre) with helping to turn his life around.

**PREFECTS**
BACK ROW: C. Foulkes, D. Pritchard, P. Anderson, S. Lill, C. Le Souef, G. Marinko, M. Edelman, A. Forrest
FRONT ROW: J. Morrison, A. Clapin, N. Aitken (Captain of School), Dr. K. Tregonning, D. Roberts, S. Lewis, S. Morris

Andrew Forrest and Jeff Braysich living the high life at Jacksons in 1987, a few months before the stockmarket crash of October that year. "You were either his best mate or worst enemy, and at various times I was both," Braysich says of Twiggy. (*The West Australian*.)

Forrest chats to John Howard at the official opening of Anaconda Nickel's Murrin Murrin nickel project in 1999. The plant was nowhere near fully operational and major technical problems would later force Twiggy out of the company and into exile. (Norm Bailey/News Corp Australia.)

Forrest in deep discussion about the Rudd government's resource super-profits tax with the public servant who recommended it, Treasury boss Ken Henry. Twiggy labelled the RSPT "economic vandalism" and played a key role in the campaign to kill it. (Ray Strange/News Corp Australia.)

Twiggy shovels the first load of Fortescue's iron ore from a wheelbarrow onto a conveyor belt at Herb Elliott Port in Port Hedland. The first shipment on 15 May 2008 meant that Fortescue had broken the Pilbara iron ore duopoly held by mining giants BHP Billiton and Rio Tinto. (Colin Murty/News Corp Australia.)

Forrest meets China's future president Xi Jinping. At the time of this meeting in Canberra, which took place on 21 June 2010, Forrest and Kevin Rudd were also attempting to broker a peace deal on the mining tax. Rudd was removed as prime minister three days later. (Mark Graham/AP Photo.)

Forrest meets Queen Elizabeth II at the Commonwealth Day service at Westminster Abbey in 2012, after addressing the audience about his philanthropic work, which signalled how far he had come in being embraced by the establishment. (Leon Neal/PA Wire.)

Forrest at the May 2010 anti-RSPT rally in Perth's Langley Park. Within weeks, Twiggy had entered confidential talks with Kevin Rudd's office with the aim of resolving the dispute over the tax. (Marie Nirme/News Corp Australia.)

Forrest shows off his horseriding skills at his Minderoo cattle station in the Pilbara. Twiggy bought the property for $12 million in 2009 after it had been out of the Forrest family's hands for more than a decade. (Ron D'Raine.)

Andrew and Nicola Forrest in 2013, after donating $1 million to the Salvation Army. Friends and family say Nicola deserves credit for instilling stronger Christian values in her husband. (Sam Ruttyn/News Corp Australia.)

Forrest with Scotty Black Jr, the son of the legendary Minderoo stockman, near Onslow in 2010. "Scotty Black was my greatest mentor outside my own blood," Forrest said. (Lincoln Baker/News Corp Australia.)

and verve had driven the Fortescue share price from 10 cents in 2003 to more than $40 in mid 2007 – a year before a single tonne of iron ore had even been shipped. Forrest became so popular that organisers at the Diggers and Dealers conference in 2007 had to lock the doors while he spoke, leaving hundreds of delegates stranded outside. The *New York Times*, the *Wall Street Journal* and the London *Times* each devoted feature articles to the buccaneering entrepreneur's exploits, with headlines like "Comeback in the outback", "Aussie jackeroo who struck it rich" and "Iron upstart challenges the big boys".

A few days after the first ship left Port Hedland, *BRW* magazine named Forrest as Australia's richest person with a fortune of just under $10 billion – at that stage the highest in the history of the annual wealth survey. It was the first time in twenty years that Kerry or James Packer had not topped the list. Forrest's wealth had grown on the list from $340 million in 2005 to $810 million in 2006 and $3.89 billion in 2007 to peak at $9.41 billion. However, it was a fortune linked directly to the iron ore price. In 2008, the price of the mineral peaked at an incredible $US200 a tonne –seven times what it had been in 2003.

Andrew Forrest had made the fastest and most spectacular fortune in Australian history. And plenty of others rode on his coat-tails. In 2003, a little-known Sydney investor, Kie Chie Wong, paid less than $1 million for a sizeable stake in Fortescue. Five years later, the Malaysian-born Wong had turned that investment into an astonishing $1.2 billion. His family company, Emichrome, had also backed Forrest at Anaconda and had thrown money at several other speculative mining stocks, but none that scaled the heights of Fortescue. Wong, from the understated beachside suburb of Maroubra, made huge profits by cashing in some of his shares over the years, but to this day he remains a top-ten shareholder with about 3 per cent of the company's shares.

Another big winner was a young Chinese metal tycoon called Wu Yueming, who turned an initial outlay of $7 million into a $700 million bonanza in just four years. After meeting Forrest at a steel conference and arranging a site visit to the Pilbara in 2004, Wu bought 7 million Fortescue shares for $1 each – a hefty 47 per cent premium on the share price at the time. As part of the deal, he also handed over $US20 million to lock in a supply deal of 4 million tonnes of iron ore a year for his Fengli Group, based 100 kilometres from Shanghai. Wu was among those celebrating with Forrest when the first batch of iron ore was unloaded in China.

Graeme Rowley, who had accumulated 19.5 million shares and become a top-twenty shareholder, was worth more than $200 million on paper at the peak of the Fortescue share price boom. Five years earlier, he'd had only $80,000 in savings to his name. Forrest's mother, Judy, also did well out of her son's success. She appeared on the Fortescue register in its early days with a parcel of 600,000 shares. By 2008 her stake was worth more than $50 million. Today she remains a top-twenty shareholder in the company. Breeding resilience in her son from an early age had helped deliver Judy Forrest a stunning windfall.

Stories also began to emerge of everyday people who'd put money into Fortescue when it was just a tiddler. For Louise O'Reilly, an Irish migrant who took a job as Forrest's personal assistant at his Cottesloe home in 2003, investing in Fortescue shares was a life-changing experience. Forrest convinced O'Reilly to put $11,500 of her savings into what was then still called Allied Mining and Processing. She sold the shares two years later for a $140,000 profit, putting the money down as a deposit on a house. It was Forrest's self-belief that had convinced her to buy the shares. "It was Andrew, basically. He is certainly a man of vision," she said.

For a brief time in 2008, Forrest's stake in Fortescue was worth a staggering $13 billion. In those days, his net worth would soar or decline by $1 billion between breakfast and lunch, depending on the vicissitudes of the stockmarket. But with Fortescue suddenly becoming a runaway success, hard questions were being asked about why local institutions had shunned the company, thereby depriving more Australians from sharing in the spoils. Even in 2007, when it looked highly likely Forrest would achieve his dream, none of the blue-chip broking houses in Australia had recommended Fortescue to their clients. The mining analysts failed miserably to foresee the approaching boom in iron ore prices and the fact that Fortescue was ideally placed to capitalise on it. In 2006, some analysts declared the run in the iron ore price to more than $US50 a tonne would not continue, yet in 2008 the price would surge to more than $US200 a tonne. In late 2007, Macquarie Equities put an "underperform" recommendation on Fortescue and a target price of $3.56. Within five months, however, the shares had climbed above $10.

It was no surprise, then, that Australian institutions accounted for a meagre 3.9 per cent of the Fortescue share register. Besides John Veldhuizen at BBY, the only real supporter of Fortescue in Australian stockbroking was Charlie Aitken of Southern Cross Equities. "The company is building the most significant new resource project in Australia, yet nobody in the Australian investment community takes it seriously," Aitken told his clients in May 2007. "Our investment community has made the mistake of playing the man and not the ball." Aitken's rivals scoffed when he predicted that shares in Fortescue, then trading at $2.50, would soar to $10 within three years. Yet he was proven right within a few months.

One of the most foolish-looking doubters was the anonymous fund manager quoted in the *Australian* in October 2007: "I'm happy to be chained to the tracks somewhere between Cloudbreak and Port Hedland in May next year because I know that I will not be killed by a train." In the Fortescue office, David Mendelawitz framed the quote and mounted it on the wall, accompanied by a picture of a fund manager in a suit standing on a railway line with the caption: "Ever wanted to flatten a banker?" It proved to be another motivating tool. Needless to say, the fund manager was nowhere to be seen when the first train hurtled down the tracks seven months later.

# 10.

# THE GREAT ESCAPE

*As Mark Twain said, reports of my death
are greatly exaggerated.*
—ANDREW FORREST

Forrest's exhilaration at proving the doubters wrong was fleeting. Within a few months of Fortescue's first shipment of iron ore, the world was gripped by the worst economic downturn since the Great Depression of the 1930s. In one sense, the global financial crisis vindicated Forrest's uncompromising push to begin exporting iron ore in May 2008, when the iron ore price was still at its peak. A delay of just four months would have meant launching the project as Lehman Brothers filed for bankruptcy in the United States, an event that triggered a severe collapse of economic confidence and a downturn in commodity prices.

When Chinese iron ore demand crashed in late 2008, Fortescue's share price came back down to earth with a thud. From a spectacular zenith of $12 in June 2008, the shares were trading below $2 by January 2009. Forrest's paper fortune of $13 billion had slumped to little more than $2 billion in just seven months.

Even in the darkest days of the global downturn, the perennially self-assured Forrest maintained that demand for commodities would bounce back quickly and that the urbanisation and industrialisation taking place in China and other Asian economies still had decades to run. In early 2009, the *Australian Financial Review* surveyed the nation's top chief executives for its half-year profit wrap and found they had little hope about the coming year. The only dissenter was Forrest, who said in his folksy style: "If you think everything is crook in Tallarook ... think again." Few knew of the turmoil at the time behind the scenes at Fortescue. Bottlenecks at the Cloudbreak and Christmas Creek mines meant production fell well behind during 2009, just when the company desperately needed to maximise revenue. As well, it was facing huge losses as a result of a disastrous foray into shipping contracts.

Fortescue was burning through cash at a worrying rate and only survived due to a $645 million equity injection from one of its key customers, state-owned Chinese steel-maker Hunan Valin, which bought 16.5 per cent of the company and became the second biggest shareholder after Forrest. After years of wrangling with Beijing, Fortescue had finally done a major funding deal with China Inc. Hunan Valin was the first Chinese steel-maker to take a large stake in an Australian iron ore miner, and Forrest hailed the agreement as an example of the economic cooperation that was needed between the two nations. Valin's enterprising chairman, Li Xiaowei, also earned a seat on the Fortescue board, although the company agreed not to increase its stake beyond 17.5 per cent, thereby assuaging concerns within the Rudd government about Chinese companies swooping on Australia's undervalued mining assets. The deal still came as a surprise to many who had long expected China's biggest steel-maker, Baosteel, to be the logical vehicle for any state-owned investment in Fortescue.

The iron ore price gradually recovered from the shock of the global financial crisis and by 2010 Fortescue's shares were again trading above $5, making Valin's purchase at $2.48 per share at the depths of the doom appear prescient. Fortescue's production problems in the Pilbara were being ironed out and the company managed to export 41 million tonnes of iron ore in 2010 to earn revenues of $3.5 billion, most of which went towards paying down debt. When Fortescue was able to make its first dividend payment to its 52,000 shareholders in 2011, it appeared to have finally arrived as a serious public company. As the biggest shareholder, Forrest received a cheque for $29 million that year and a further $80 million in dividends over the next twelve months.

On 18 July 2011, exactly eight years after founding the company, Forrest stood down as chief executive and handed over to Nev Power, a physically imposing former Thiess executive from Queensland whom Twiggy had recruited after years of searching for the right person to succeed him. Forrest declared he wanted to spend more time on his philanthropy, denying that his decision to resign as chief executive was linked to his ongoing battle with the Australian Securities and Investments Commission over allegations he had misled the market.

Nev Power shared Forrest's passion for overly ambitious targets, but he also boasted the operational expertise that the company had badly needed ever since it began exporting iron ore. Forrest had proven to be a brilliant fundraiser and a visionary leader, but his time at Anaconda had shown that running an established company was not his speciality. Soon after announcing his retirement as chief executive, Forrest even admitted at a business conference that his day-to-day management skills were lacking: "I think there are some Fortescue Metals people here in the audience that will shake their heads if I say I'm a good manager. I'm not such a great manager." Despite Forrest's move into the background, Nev Power was never going to become the sole public face of Fortescue. Forrest stayed on as non-executive chairman and still owned a controlling 32 per cent of the company, ensuring he had the ultimate say on all the big issues.

With iron ore prices creeping back towards $US200 a tonne during 2011, the value of Forrest's stake was back up around

$6 billion for much of the year, ensuring he remained near the apex of the *BRW* Rich List. But this was when hubris began to creep in. Several months earlier, Forrest had unveiled an extraordinary target to boost Fortescue's market value to at least $100 billion and to expand annual iron ore production sixfold, to 355 million tonnes by 2017. To put these ambitions into perspective, even Rio Tinto – the biggest iron ore producer in the Pilbara – wasn't planning to be producing that much iron ore by then. "Forrest only chose that number [355 million] because it was 5 million more tonnes a year than Rio," says a former Fortescue executive. At the time, Fortescue had not even hit its initial target of producing 55 million tonnes a year. Analysts began to worry that Forrest, who for his whole career had wanted to be the biggest and the best, was getting ahead of himself. "Right now the focus needs to be about delivering the first 55 [million tonnes]," said a sceptical UBS analyst, Glyn Lawcock. "Let's take it one step at a time."

By 2012, economic growth in China was beginning to slow down and iron ore prices were retreating from their unsustainable peaks of the previous year. A well-known American hedge fund manager and short-seller, Jim Chanos, began telling his clients that he was "shorting" Fortescue shares, or betting that the price of the stock would fall, because the miner was too highly geared, too exposed to China and was being led by a "promotional management team". Although Forrest had handed over management to Power by then, the barb about self-promotion was aimed directly at him.

Chanos also predicted that Fortescue would have trouble servicing its huge debt pile if iron ore prices fell below $US100 a tonne. At the time, such a prospect appeared unlikely because iron ore was trading steadily at around $US150 a tonne, allowing Fortescue to meet its interest payments and still deliver a healthy return for shareholders. Moreover, Fortescue's share price had risen 30 per cent between January and April, despite concerns about the strength of China's economy. Both Forrest and Power went in hard against Chanos, suggesting Fortescue would always be protected by a natural "floor" in the iron ore price of $US120 a tonne.

Within a few months, however, almost everything Chanos had warned about Fortescue appeared to be coming true. China's economic growth rate began to slow further and the iron ore price spiralled below the so-called floor of $US120 a tonne. For Fortescue, it was the worst possible timing: the miner's revenues collapsed at the same time as its debt load peaked. The debt was funding most of the company's $US8.4 billion plan to almost triple its production capacity in the Pilbara to 155 million tonnes a year by mid 2013 (Forrest's previous extravagant target of 355 million tonnes by 2017 had by then been placed on the backburner).

Forrest had long rejected the idea – even when raised by his fellow directors – of raising additional equity rather than taking on even more debt, because this would have diluted his controlling 32 per cent stake in Fortescue. An equity raising now made even less sense for him, given the sudden fall in the share price. Since the early days, when he recruited his inaugural directors, Forrest had ruled the Fortescue board with an iron fist. But tensions developed when Americans Ian Cumming and Joseph Steinberg, the Leucadia bosses who both served stints as non-executive directors of Fortescue, grew critical of what they saw as Forrest's aggressive growth strategy and his "severe exaggeration". They told Leucadia shareholders in 2013: "His personality dominated the FMG board and the other directors were more inclined to follow his lead as to the appropriate amount of equity, debt, leverage and the rate at which to expand, as opposed to our more conservative views."

Ratings agency Moody's didn't help matters by putting Fortescue's credit rating on review for a possible downgrade in August 2012, leading to further share price falls. Forrest tried to outsmart the short-sellers by borrowing money to buy $175 million worth of stock himself. At last, he had some real "skin in the game", taking a personal financial gamble on the share price improving. But his extraordinary outlay failed to halt the decline. He next embarked on a solo mission to shore up confidence in his company by talking up its prospects to anyone who would listen. "I think it's a great investment, it's a multi-generational company, it will be going for many decades and will be one of Australia's great institutions and I feel privileged to own shares in it," Forrest said with habitual buoyancy. He added that Fortescue was in "great shape" and could "weather any storm".

But when the iron ore price dropped like a stone to $US87 a tonne in early September, Forrest had egg on his face. At that price, Fortescue was simply not profitable. The iron ore divisions of BHP and Rio were not facing anything like Fortescue's conundrum because they had been funding their Pilbara expansions from their own balance sheets rather than by taking on debt. Fortescue's costs of production were also far higher than either of its Pilbara rivals and its ore was of generally lower quality, meaning its profit margins were much thinner.

Rumours began to swirl through the market that Fortescue might go under. The sceptics who had never trusted Twiggy quickly re-emerged to point out the parallels between Fortescue's woes and Forrest's troubled period at Anaconda Nickel – particularly the reliance on big slabs of debt and a deep exposure to a single commodity. The short-sellers, in particular, were triumphant at what they saw as the clear shortcomings of Forrest's business model. One of them, John Hempton from Bronte Capital in Sydney, summed up the feeling in an interview with Reuters: "He [Forrest] builds iron ore mines on a vast scale. Then he borrows money on a vast scale. And he presumes the iron ore price will deliver him a pile of loot that is also unimaginable to mortals. The world is not complying."

Faced with a looming disaster, Forrest and Power took drastic action on the morning of Tuesday 4 September 2012. Over the previous weekend, dozens of staff members had been called into head office to work out how they could urgently slash costs. The senior executive team met in the morning and agreed to come back at regular intervals until 8pm to discuss the latest drafts of their cost-cutting plan. Workers munched on sandwiches from a nearby Subway outlet as well as sweets left over from the eighteenth birthday party of chief financial officer Stephen Pearce's daughter the previous evening. Forrest, Power and Pearce decided that their debt-funded expansion plan – which they had repeatedly promised would be unaffected by any iron ore price collapse – would need to be radically scaled back, and that 1000 employees and contractors would have be axed if the company was to survive.

By that afternoon, hundreds of vacant desks at Fortescue's head office provided vivid evidence of the carnage. Entire teams, including the environmental, heritage, legal, IT and exploration divisions, were cleaned out. Many of those working in the company's indigenous relations group were also sacked, in spite of Forrest's commitment to improving living conditions for Aboriginal people. Fortescue said the retrenchments would save $300 million in operating costs that year.

After all of Forrest's preaching over the years about the love he felt for his "Fortescue family", the company's actions that day confirmed that the quest for profits would always come before any sense of loyalty. One worker posted a note on Facebook claiming his email account and phone were disabled as he was being escorted from the Fortescue building. "I couldn't even say goodbye to those I wanted to in the office," he wrote.

The biggest shocks were still to come, however, when several of Fortescue's most respected and longest-serving senior executives were culled in the following days. Years of loyalty to Forrest and his dream suddenly counted for nothing. Those sacked included government relations chief Julian Tapp, the key strategist behind Fortescue's lengthy efforts to gain access to BHP's railways and a major figure behind Forrest's campaign against the federal government's mining tax. Also axed was investor relations chief Rod Campbell, who had been sent out the previous day to brief analysts on Fortescue's pared-back expansion plans.

Tapp and Campbell, who both joined Fortescue in 2004, were blindsided by their sackings. Campbell and his wife had enjoyed a close relationship with Andrew and Nicola Forrest over the years. Like others, they struggled to understand how the decision to remove them and pay them up to six months' salary in redundancy packages could benefit a company facing a short-term cash squeeze.

Another senior executive, Ann Marie Lowry, had accompanied Forrest to Sydney on 4 September as he made an impassioned speech at a Philanthropy Australia conference about the evils of child labour. Lowry, the company's senior legal counsel, was known as the "chief cheerleader" for Fortescue and had grown personally close to both Andrew and Nicola. In an in-house publication in 2008, she said: "Of all our assets, the ones we value most and see as most powerful are our people and our culture." But by the end of the day, even as she travelled with the Forrests, Lowry was sacked.

Geologist Barry Knight, who had performed a variety of senior roles since joining Fortescue in 2003, when it was still housed in Forrest's lounge room, was also retrenched. Eamon Hannon, the once-fêted exploration boss whom Forrest treated as a son, was so disgusted at the way some of his staff were retrenched that he brought forward his own planned resignation date and walked out the door. One long-serving former executive says there is still anger towards Forrest and Power over the sackings, which he believes were badly handled. "Like at other companies, it should have been last in, first out," he says. "It was a very sad day for Fortescue."

While many saw the purge as an example of Power trying to step out of Forrest's shadow and put his stamp on the company, Forrest admitted later that he had personally endorsed all of the sackings. But he did not personally deliver the bad news to anyone; that grim task was handled by Power and Stephen Pearce. When asked about the sacking of some of his most loyal executives, Forrest said he could not countenance a situation in which those with close links to him were excluded from redundancies that were needed to keep the company afloat.

Just a few days later, he said he was confident that Fortescue would emerge from its scrape with death in stronger shape. "I can't tell you how well I am sleeping at night," he said. "The share price thing, it really matters nought to me. But my heart breaks for the people we have had to let go. They are paying a high personal price for something that is necessary because we have to emerge from this with a company that is fitter and stronger to make sure we really have built a multi-generational business."

Under the scaled-back growth plan, Fortescue's new annual production target would be cut to 115 million tonnes – a move that would save $1.6 billion for the year. Power also announced a fire sale of some of the company's "non-core" assets. He managed to sell a power station to a Canadian utility for $300 million and a 25 per cent interest in an iron ore joint venture with fellow miner BC Iron for $190 million. In late 2012, Fortescue also started talks aimed at selling a stake in its Pilbara rail and port assets, a move that could raise billions of dollars.

But the bean counters at Fortescue's head office in Perth had even further cost-cutting in mind. On 5 September, an internal memo announced that staff barbecues would be axed in Port Hedland. The memo also said the company would no longer buy tomato sauce or other condiments for staff, who would also have to bring their own cutlery from home in future. And for good measure, employees were warned that the stationery cupboard in the administration office would from now on be locked. Morale among Fortescue staff hit rock bottom.

Largely unbeknown to the market, Fortescue was facing a separate financial problem that posed a much bigger threat than a few bottles of ketchup. The collapse in the iron ore price had rendered Fortescue unprofitable, but it also meant the company would likely be in breach of its loan covenants before the year was out. The conditions on the loans required Fortescue's pre-tax earnings to remain at more than two and a half times the annual interest bill on its $US10-billion debt. On 13 September, the *Australian Financial Review* published a story on its website that revealed Fortescue was in talks with its bankers about a possible refinancing. The story triggered a 14 per cent plunge in the share price to just $2.99 and yet another round of speculation that the company was facing collapse.

In Perth, where Forrest is a household name, even taxi drivers were suggesting the tycoon's luck had finally run out. The main story on the front page of the *West Australian* the next day suggested Fortescue was in danger of going under. Like Icarus, Forrest appeared to have flown too close to the sun and was now crashing back to earth. The website *Crikey* ran a story gleefully reminding Forrest how he had ignored Jim Chanos' warning about the twin dangers of Fortescue's high gearing levels and the falling iron ore prices. "The disaster unfolding at Fortescue, and the desperate attempt by billionaire Andrew 'Twiggy' Forrest to cling on to his fortune, prove again that when Jim Chanos says something, it's worth listening," wrote business reporter Adam Schwab.

Mere hours after those words were published, Fortescue announced it had raised $US4.5 billion in fresh debt to shore up its sickly balance sheet. Power and Pearce endured sleepless nights to complete the deal, while Forrest travelled to China, the United States, the Middle East and Europe, talking to shareholders and financiers. The new facility would refinance all of Fortescue's existing bank debt and remove the restrictive covenants that would otherwise have been breached within months. In short, Fortescue had taken on more debt to solve its debt crisis, but the fresh loans came with less onerous conditions and no principal repayments for three years. The company had bought more time in the hope the iron ore price would recover. It was a sign that credit markets still had a taste for risk and continued to believe in Forrest's vision. Just a few days after many had written him off, Twiggy had pulled a rabbit from his hat.

In raising the money, Fortescue was also able to walk away from its 2006 loan deal with Leucadia, which had ended up costing the miner more than $1 billion in royalty payments. In 2010, Forrest hired prominent Melbourne investment banker John Wylie and former Labor prime minister Paul Keating, an adviser to Wylie's firm Lazard, to lobby Leucadia to extinguish the loan. Over the years, Leucadia boss Ian Cumming had revelled in the brilliance of his loan deal with Forrest, once describing it as "succulent". Because the royalty payments under the loan were linked to Fortescue's iron ore revenues, the deal had proven far more profitable for Leucadia than even the most bullish forecaster could have predicted at the time.

On their initial mission to New York in 2010, Wylie and Keating failed to convince Cumming, a 71-year-old billionaire, to give up the loan. It didn't help matters that Leucadia was also suing Forrest in the WA Supreme Court, alleging he had engaged in "misleading and deceptive behaviour" in attempting to dilute the hedge fund's share of future profits from the convertible note. Leucadia sold down most of its shareholding in Fortescue in 2011, but Cumming's anger with

Forrest was doubtless assuaged by the fact that his original equity investment had delivered a profit of more than $1 billion.

By September 2012, with the iron ore price in freefall, Wylie went back to Forrest and said he believed it was an ideal time to negotiate with Leucadia to buy it out of the loan, which was not due to expire until 2019. According to the *Australian Financial Review,* Wylie met Cumming in the United States and said Fortescue was willing to pay $US715 million for the note if Leucadia dropped its legal action in Australia. He warned that Fortescue was about to unveil a huge secured debt facility that would put Leucadia towards the bottom of the list of creditors in the event of the company being wound up – a scenario that wasn't unrealistic at the time. Cumming accepted the offer. What neither Wylie nor Forrest knew at the time was that Leucadia urgently needed cash to fund a major acquisition it was planning. Just as the Leucadia deal had proved to be brilliant timing for Fortescue back in 2006, its ability to extract itself from the expensive loan seven years early was also something of a masterstroke.

In the space of one wild month, Fortescue had appeared close to financial ruin, only to emerge with a new financing package and free from the Leucadia loan. But in reflecting on the drama, the *Australian*'s respected business commentator John Durie pinned the blame for the near-death experience on Forrest's hubris in pushing to expand Fortescue at such a rapid pace. "Andrew Forrest is financially secure once more, but questions remain whether he has at last learned the perils of too much debt in a commodities play," Durie wrote. "The way Forrest saw the world, the iron price would stay high forever and the quicker he expanded the quicker he could lower the costs of production and hence make more money."

Of course, Forrest's expand-at-all-costs mentality and his penchant for risk were not new. He had done the same thing, with varying degrees of success, in deal-making circles in Perth and Sydney in the 1980s and 1990s. Similarly, his grandiose ambition to dominate the nickel industry at Anaconda in the 2000s had proved unrealistic. But Forrest thrives on dealing in high stakes and proving people wrong. As his executives scrambled to refinance Fortescue's debt in the darkest days of the price plunge in September, a photograph of a beaming Forrest appeared on the front page of the *Australian Financial Review*, accompanied by a story in which he declared that suggestions of his imminent failure were not credible. "As Mark Twain said, reports of my death are greatly exaggerated," he said.

Forrest's blind optimism underscores legendary British economist John Maynard Keynes's theory that business decisions are often driven by "animal spirits" rather than a careful assessment of the numbers or the facts. Keynes suggested that without "animal spirits", or entrepreneurs investing spontaneously, there would be no economic progress. "It is safe to say that enterprise which depends on hopes stretching into the future benefits the community as a whole," he wrote. "But individual initiative will only be adequate when reasonable calculation is supplemented and supported by animal spirits, so that the thought of ultimate loss which often overtakes pioneers, as experience undoubtedly tells us and them, is put aside as a healthy man puts aside the expectation of death."

There is a big difference, though, between animal spirits and recklessness, which is how some have viewed elements of Forrest's career. The 2012 debt crisis exposed Fortescue's core weakness: it is entirely reliant on strong iron ore prices and it is entirely funded with other people's money. Forrest has also failed to diversify Fortescue beyond iron ore, despite promises over the years to push into coal and other minerals. Philip Kirchlechner, who served as Fortescue's head of marketing between 2003 and 2006, believes Forrest made a critical error in failing to encourage the Japanese or Korean steel industries to buy his iron ore. "Forrest is hugely exposed to a single market in China," he says. "BHP and Rio are very, very jealously guarding their sales into Japan and Korea – those markets are important to them in terms of diversification. But Forrest always had trouble in dealing with Japan because the Japanese couldn't deal with his aggression. It was okay for China but not in Japan."

When the debt refinancing deal was announced, the Fortescue share price surged 17 per cent in a single day. But analysts warned that Fortescue remained too highly geared and exposed to the highly volatile iron ore price. They wondered whether Forrest's great gamble would survive the next crisis. Many also cautioned that the iron ore market would suffer from a significant oversupply after 2014, by which time prices could be back down below $US90 a tonne. The analysts, of course, had been proven wrong before. And in Forrest's world, they would be wrong again.

Proving that life at Fortescue was indeed a roller-coaster ride, the iron ore price climbed back to $US135 a tonne by December 2012 and the company was able to reactivate the $1.1 billion development of its Kings deposit, which would deliver an extra 40 million tonnes a year of iron ore. The production target of 155 million tonnes a year was back on the agenda just a few months after it had been shelved. Fortescue appeared to be back on track, albeit with a $12.8 billion debt pile, which was higher than the company's market value.

Despite this semblance of stability, Power was adamant that there was no place for the 1000 people axed in September, conceding that Fortescue had grown fat in the good times and would have to remain a lean organisation. It was a very different message to the one Forrest had delivered in Fortescue's 2012 annual report, when he declared that all former servants of the company would "always be welcome back" when the opportunity arose.

# 11.

## CHINA CRISIS

*Sometimes you've got to do things to survive
that you would have preferred not to do.*
—Rodney Adler

From the moment Forrest burst back on to the scene in 2003, the nation's corporate cops began to keep an abnormally close eye on him. The regulators at the Australian Securities and Investments Commission (ASIC) and the Australian Securities Exchange (ASX) – the two bodies tasked with protecting investors – were wary of Forrest, who was intent on winning publicity for his new venture. They also knew investors had lost plenty of money at Anaconda Nickel when Forrest had promised the world but failed to deliver. And besides, they had a long-standing mistrust of big-talking mining entrepreneurs from Perth.

Ultimately, these suspicions would lead to an eight-year legal battle over Forrest's conduct that would jeopardise his ability to remain in charge of Fortescue, cast a pall over his reputation and cause untold stress to him and his family.

The regulators' scrutiny of Forrest's grand vision for Fortescue began within days of his election as chairman in July 2003. Perhaps the ASX was uneasy that Fortescue's share price had risen rapidly from 10 cents in April to 26.5 cents on the day of the July shareholders' meeting, even though there had been no official announcements about Forrest's planned takeover of the company.

At the meeting at the Celtic Club in West Perth, Forrest could hardly have been accused of talking down his prospects of success in building an iron ore mine in record time. He announced he was seeking a partnership with mining heiress Gina Rinehart's planned Hope Downs iron ore project in a bid to establish new rail and port facilities in the Pilbara. But a Hope Downs source was quoted in the *West Australian* three days later as saying Forrest's infrastructure plan was "not something we are spending a lot of time thinking about at this stage".

Fortescue quickly released what it termed an "explanatory statement" to Forrest's speech in an attempt "to ensure there is no confusion in the market". The statement confirmed Fortescue had no agreement with Rinehart's Hope Downs project. It went on to "clarify" to the ASX that Fortescue's proposed iron ore project had no definitive reserves, no definitive budget, no definitive timetable and no definitive funding strategy.

On the same day, the ASX sent Fortescue a "please explain" letter over another report, which said the company was in talks to form an iron ore joint venture with fellow Perth company Consolidated Minerals, a move that would boost Fortescue's chances of being able to raise the money it needed for a railway line and a port. The ASX clearly wasn't happy that it was reading about the talks in a newspaper before the market had been informed. Later that day, Fortescue confirmed the joint venture with Consolidated Minerals had been finalised.

These early misunderstandings would seem trivial compared to the stoushes that followed. In December 2003, a team of Fortescue executives, including Forrest, Graeme Rowley, Alan Watling, Chris Catlow, Barry Knight and Philip Kirchlechner, embarked on a crucial trip to the Chinese industrial city of Maanshan to spruik their vision to thirty of the country's steel mills. As a result of the trip, Fortescue announced on 9 December it had secured "memoranda of intent" with steel mills to buy 25 million tonnes a year of iron ore, declaring it was confident it had established a market for its product. But the ASX put a dampener on the news when it asked Fortescue why its share price had suddenly run up from 48.5 cents to 66 cents in the four trading days before the announcement of the MOIs.

Among the buyers during that period was Graeme Rowley, the Fortescue executive director who was also Forrest's right-hand man. On 3 December, before the Maanshan trip, Rowley had paid $170,000 to buy 350,000 Fortescue shares at 48.5 cents each. By 9 December, when the Maanshan deals were announced to the market, Rowley's $170,000 outlay was suddenly worth $230,000. ASIC wanted answers and summoned all of the Fortescue executives who went to Maanshan in an attempt to discover what Rowley had known about the status of the MOIs when he bought his shares.

ASIC told Fortescue it was investigating Rowley under section 1043A of the Corporations Act, otherwise known as the insider trading provision. The Fortescue executives, including Forrest, gave sworn evidence at ASIC's Perth office and were told to hand over all relevant documents. "Graeme was stupid to buy shares at that time," says a former Fortescue executive. "I think it was an honest mistake, but I still thought it was stupid."

ASIC dropped its insider trading investigation into Rowley several months later and the probe was never made public, but the episode illustrated that the corporate regulator had Fortescue firmly in its sights from the early days. Later, ASIC also questioned Fortescue over the nature of some of the "binding" sales contracts it had signed with Chinese steel mills in late 2004 and early 2005. In the end, that investigation was also dropped.

But the regulators stayed on Forrest's case. In April, the ASX sent an extraordinary four-page letter to Fortescue, demanding answers to a series of questions about the company's project cost estimates, its timelines and the extent and quality of the mineral resources it had announced to investors. This was a rare kind of rebuke, and it proved that Fortescue was being scrutinised more closely than just about any other stock on the market.

The questioning led Fortescue to admit that its project was indeed facing a significant cost blowout beyond the budget of $1.85 billion and a six-month delay until the second quarter of 2007. It was also forced to back down on the bullish way it had been describing its resources to the market over the previous eight months. The ASX said Fortescue's reporting did not comply with the Joint Ore Reserves Committee (JORC) code, which is the official measure by which miners must evaluate their deposits. It pointed out that Fortescue should not be using the term "ore" to describe its mineral resources because that implied it had a "reserve" that was viable for mining. The company's response shocked the geologists at other mining companies, who viewed the industry's JORC code as sacred. "It was not Fortescue's intention to use the term 'ore' in the context of suggesting that it was at 'reserve' grade status," Fortescue said, "but rather the term was being used in its colloquial or natural meaning within the 'iron ore' industry."

Within a few days, there was more to come. The perennially bullish Forrest, his popularity in mining circles clearly rising, gave a sold-out address to the Sydney Mining Club in May 2005, in which he said Fortescue would be able to sell its lower-quality ore for 95 per cent of the prices being obtained by BHP Billiton and Rio Tinto for their premium products – a claim that raised eyebrows when reported in the *Sydney Morning Herald* the next day. Forrest also told his audience that the capital cost of the project "would start with a two, not a one" but he would be working hard to ensure it did not blow out to anywhere near $3 billion.

The ASX jumped on the statements, demanding the company explain what Forrest had meant. Fortescue responded that everything Forrest had said was correct. This may have been true, but to many observers it appeared as if Fortescue had a real problem in complying with Australia's system of continuous disclosure, which requires companies to announce all price-sensitive information to the market in a timely manner.

It was in March 2006, however, that Forrest felt the full weight of the nation's regulators. After a twelve-month investigation that arose from an ASX referral, ASIC's then chairman, Jeff Lucy, announced the watchdog was suing Forrest and Fortescue in the Federal Court. ASIC alleged Forrest had breached his duties as a director by referring to a series of construction deals with Chinese companies as "binding contracts" when they were, in fact, merely loose framework agreements. It also claimed Forrest had failed to correct his misleading statements even when it became obvious that the deals were not binding. In essence, ASIC was saying Forrest had exaggerated the nature of the deals and then deliberately kept the market misinformed while Fortescue's share price soared. It wanted Forrest to be fined $4 million and, more importantly, disqualified from serving as a director over what it regarded as a serious breach of the Corporations Act. Experts said the ban would likely last for between three and five years if he were found guilty.

For Forrest, the stakes had never been higher. To his critics, the civil charges were confirmation of what they believed all along: the Forrest spin machine had finally gone too far and he would be brought to account by a judicial system that had already found him to be dishonest several times in the past. But to Forrest's growing band of supporters, ASIC's pursuit was evidence of Australia's "tall poppy syndrome" and the regulator's desperate need to claim the scalp of a high-profile businessman. Forrest himself responded by saying he had done nothing wrong, the contracts were indeed "binding" and he was confident the courts would find in his favour.

Another school of thought contended that even if Forrest had overstated the nature of the deals to generate some excitement and boost his share price, a protracted courtroom brawl involving arcane legal argument over the precise definition of a "binding contract" might not be the best way to remedy the breach.

Forrest suspected that the ASIC charges had arisen as a result of lobbying by his enemies at BHP and Rio. There was no direct evidence to support this. But the damage to Forrest's reputation was immediate and it allowed his rivals in the industry to privately gloat over the whole affair. In the *Australian,* business columnist Matthew Stevens called on Forrest to resign from Fortescue, arguing that the charges could jeopardise the company's ability to secure contracts with financiers and customers.

The trouble had begun in August 2004, when Fortescue had told the market it had entered into a "binding contract" with the state-owned China Railway Engineering Corporation (CREC) to finance and build its railway line in the Pilbara. At a press conference, a journalist had asked Forrest: "You talk about a $1.85 billion project, how much of that is the railway line?" Forrest had replied: "The price of the railway line and the rolling stock is confidential but we are pleased to say it's competitive." To the average investor, it sounded as if Fortescue had already negotiated a price for the works. But Forrest was too busy celebrating to worry about such semantics; that evening he was marking the deal by downing cocktails at a lavish party at the Australian Embassy in Beijing, surrounded by a throng of senior Chinese government officials and businessmen. Ten weeks later, Forrest had even more good news to trumpet: Fortescue had signed "binding contracts" with two other Chinese state-owned entities, China Harbour Engineering Company (CHEC) and China Metallurgical Construction Corporation (MCC), to finance and build its port and mine.

The mining industry had gasped in astonishment at these announcements. It seemed as if Forrest might really have a viable project. Forrest claimed the deals had removed the risk of the project. "These commitments by Chinese interests now cover the financing and construction risk for the total project," he said after a signing ceremony with CHEC and MCC in Beijing. "Our approach has been to ensure the construction risk is carried by the contractors and that project payment by Fortescue Metals only follows practical completion." In the *Weekend Australian*, resources writer Robin Bromby had summed up the mood among observers by declaring: "Andrew Forrest has pulled off what must be one of the most breathtaking deals in our mining history – he has talked the Chinese government into almost fully financing his $1.85 billion Pilbara iron ore dream. And he has done it without giving away a single share in his Fortescue Metals Group."

Investors clearly thought the same thing. Fortescue shares were trading at just 55 cents when the CREC announcement was made on 23 August. By the time the latter two "binding contracts" were announced in November, the share price had risen to $1.66. From there, the shares continued to soar, rising above $5 in February 2006 and catapulting the company into the ASX top 300. Fortescue had arrived as a glamour stock.

But as investors rushed to get behind Forrest's vision, his so-called binding contracts were beginning to unravel behind the scenes. In the early weeks of 2005, the Chinese companies began to demand a big equity stake in the project as a condition of going ahead with the financing deals. But Forrest was not prepared to give away majority control of his company or the project.

According to evidence given in the Federal Court, Forrest embarked on a desperate campaign during early 2005 to get the Chinese to commit to the contracts. "We have no money. We might have to think about a joint venture," Forrest was claimed to have said. He was even apparently prepared to invoke his friendship with prime minister John Howard in an attempt to force the Chinese to honour the contracts.

In mid January, Forrest met MCC vice chairman Ma Yanli, who told him he was reluctant to commit to the project before Fortescue had proven it had sufficient iron ore reserves. A Fortescue translator and executive assistant, Wei Fisher, told the court that Forrest had said to Ma during the meeting: "This is a very important project and MCC is a large company. I have been to see the prime minister, and he knows that there is a signed agreement. If MCC pulls out now, the ability of the Chinese to carry out international work will be doubtful and it would damage our relationship."

On 3 February, as his net worth soared on the back of the Fortescue share price frenzy, Forrest was handed a fax that had just arrived from Ma. According to evidence given in court (but which was ultimately rejected by the judge), the fax said the deal MCC had signed three months earlier to build Fortescue's mine was off. Wei Fisher, who was standing near the fax machine at the time, told the Federal Court: "I recall that when Andrew Forrest saw this fax, I heard him say words to the effect that 'This is ridiculous. If anyone tells the press, we have had it.' I recall seeing Forrest walking about the office waving the fax, and raising his voice, he was very upset."

According to Fisher, Forrest then went into a meeting with Graeme Rowley, Fortescue's chief financial officer Chris Catlow and legal adviser Peter Huston. "After lunch I was called into Graeme Rowley's office at about 3pm," she said in her witness statement. "He said words to the effect: 'You are no longer needed by FMG. You haven't done the job that you were expected to do. You did not do well in the negotiations with the Chinese.'" Fisher said she was upset at being sacked and tried to raise the issue with Forrest, who simply said: "This is Graeme's decision."

Forrest perhaps should not have been shocked that the Chinese were attempting to back out of the deal, because a similar thing had happened to him two years earlier. On that occasion, Forrest had signed a memorandum of understanding in Perth with another state-owned company, China Railway Materials, which had agreed to inject $20 million into Fortescue. Forrest, his senior executives and the management of the Beijing company, including vice president Liu Guoping, gathered to drink endless rounds of the Chinese hard liquor *baiju* to celebrate the deal at the Jade Court restaurant in Cottesloe. But Liu was forced to abandon the deal – which was never made public – just a few weeks later, when his masters in Beijing refused to endorse it. It was a warning of the potential hazards of doing business with China.

In March 2005, a reporter with the *Australian Financial Review* in Shanghai, Stephen Wyatt, interviewed MCC's president, Shen Heting, about the company's agreement to build Fortescue's mine in the Pilbara. Shen told Wyatt that he did not believe that MCC, which was leading the negotiations on behalf of the Chinese, had a binding contract and that none of the construction companies was prepared to finance the project unless they had a better idea of the extent of Fortescue's iron ore resources. He added that the Chinese groups would not pursue the project unless they could acquire a majority equity interest in it. Wyatt knew he had a big news story on his hands and moved to publish it before the upcoming Easter long weekend.

Wyatt's explosive story, published on Thursday 24 March, the day before Good Friday, was accompanied by an analysis by the paper's resources editor, Ian Howarth, who pointed out what BHP, Rio Tinto and Forrest's other enemies had been saying for a long time:

> The ability of Mr Forrest, a persuasive marketer, to charm investors with the company's "story" has helped propel the stock more than 700 per cent over the past twelve months alone. But in reality Fortescue amounts to nothing more than

a $1 billion concept stock. Despite peppering the Australian Stock Exchange on an almost daily basis with announcements about the company's progress, there are no mines, no railway lines, no ports and not one piece of iron ore has been sent to Asian customers.

Predictably, once the story was published, the ASX demanded immediate answers from Fortescue about the nature of its contracts. The company responded after the Easter break by releasing copies of the three deals to the market. But the documents showed that the "binding contracts" were all headed "framework agreements" and contained no reference to cost, the scope of the works, scheduling or any other details that would normally be included in a contract enforceable under Australian law. Each of the agreements ran to only four pages. In spite of this, Fortescue insisted the deals still obliged the Chinese companies to build the infrastructure. To ASIC, however, there was nothing legally binding about them. The regulator believed they were little more than loose agreements between the parties to keep talking in the hope of working out all the critical details at a later date.

The market responded to the release of the contracts by sending Fortescue's share price down a colossal 25 per cent in a single day. It would be nine months before the shares would trade at the level they had been when the *AFR* article was published. Investors clearly did not agree with Forrest's assertion that he had binding contracts with the Chinese to build the infrastructure.

Forrest was facing the prospect of being banned as a company director, but for many anxious weeks he faced something even more serious: the prospect that ASIC would also pursue a criminal prosecution against him for alleged insider trading in Fortescue shares. When ASIC filed its original statement of claim with the Federal Court in 2006, it dropped in the salacious detail that Forrest had sold $13.5 million worth of shares in February 2005 at the very time the market was allegedly misinformed about the nature of the Chinese deals. Forrest had sold the stock on Valentine's Day at a time when the share price was rising rapidly, thanks to the so-called binding contracts. The *Australian* and the *Age* reported that ASIC was weighing up whether to lay charges against Forrest and would make a decision by the time its main case reached court for a directions hearing. But ASIC never pursued a criminal prosecution.

Another uncomfortable revelation was that Forrest's friend, and Fortescue deputy chairman, Herb Elliott had also sold more than $500,000 worth of shares just days before the price plummeted in late March over the revelations about the China deals. Documents lodged with the ASX showed that Elliott had sold 100,000 of his 900,000 Fortescue shares on 18 March 2005 – six days before the *AFR* article was published. Elliott told the *West Australian* he had no idea about the emerging doubts over the Chinese contracts at the time he sold the shares. "Nobody knew anything about it," he said. "Nobody in the world did. Apart from the Chinese bloke who made the statement."

By the time ASIC's case against Fortescue and Forrest reached the Federal Court for a full hearing in 2009, Fortescue had become an established iron ore exporter and Forrest was a billionaire with the means to hire the best lawyers in the land. Four years earlier, when the investigation was launched, Fortescue was just a speculative company with a market value of less than $100 million and Forrest was anything but a household name. Still, neither side wanted the expense of a protracted courtroom brawl. ASIC and Fortescue negotiated with the aim of settling the case before it reached trial, but could not reach an agreement. At one stage, Fortescue had offered to make a payment as long as no admission of wrongdoing was recorded. But that wasn't enough for ASIC, which was convinced it had an open-and-shut case.

On the seventh floor of Perth's grey Federal Court building, a gaggle of lawyers lined up before Justice John Gilmour to begin an epic battle that would take years to resolve. In Forrest's corner was the brilliant Melbourne lawyer Allan Myers QC, himself one of Australia's richest people with a fortune in the hundreds of millions of dollars, thanks to a canny investment in the Polish brewing industry. Myers charges $20,000 a day, but few of the wealthy business figures he has represented over the years have doubted the worth of his services.

Representing Fortescue at the hearing was another Melbourne barrister, John Karkar QC, whose success in a number of high-profile cases allowed him to charge almost as much as Myers. ASIC was represented by a former Federal Court judge, Neil Young QC, who as a barrister had nailed several big-name defendants, including Steve Vizard and John Elliott. Forrest did not turn up to court to listen to any of the evidence, and he chose not to enter the witness box himself. He was, however, interviewed under oath by ASIC and chose not to claim the privilege against self-incrimination.

In his opening address, Karkar attempted to paint the dispute over the contracts as a simple clash between Australian and Chinese business cultures. He described the *Australian Financial Review* story as "a pack of lies" that had been sparked by Beijing's anger at being refused a controlling equity stake in Fortescue. "It was, to use neutral language, a negotiating ploy," he said. "For the *Australian Financial Review* to take it up was misjudged." Karkar said the key middleman in the deal between Fortescue and China, government official He Lianzhong, had promised to "teach them [Fortescue] a lesson" after Forrest had refused to let MCC take a stake. (At the time of the trial, He, who was in charge of outward foreign investment at the powerful National Development and Reform Commission, had actually just been sentenced to twelve years' jail for receiving more than 1 million yuan in bribes; the conviction was unrelated to his dealings with Fortescue.)

When it was Allan Myers' turn to speak, he didn't mince words. He said investors in Fortescue would be "weeping tears of joy" even if they had been misled. No investor had complained about Forrest's behaviour and none had suffered any loss. "ASIC's case is an attack on my client, who … took a company that, shortly before these events, was conducted from his living room to the ASX 300 by March 2005," Myers said. "This whole pleading is like a big soufflé. There's a little bit of egg white and a little bit of batter and it's whipped up a with a lot of hot air to make it bigger than it is." Myers said Forrest had not breached his duties as a director and he was entitled to use the "business judgment rule" of the Corporations Act as a defence, which gives protection to a director if a decision is made in good faith.

Neil Young, for ASIC, pointed out that the regulator was not obliged to show that investors had suffered any loss to prove its case that the statements were misleading. He also argued that Forrest must have known that the agreements with the Chinese could not be legally binding until the equity issue had been sorted out.

When ASIC began to present evidence of Forrest's alleged misdemeanours, its case quickly began to fall apart. The regulator's star witness was Ed Heyting, the disgruntled former Fortescue executive who had drafted the contracts. ASIC had expected Heyting to confirm in forthright language that the contracts he drafted were never intended to be legally binding. But under fiery cross-examination from Karkar, he was forced to back down on key elements of his witness statement. His stint in the witness box ended up working against ASIC, leading Justice Gilmour to find his evidence "far from persuasive". Other ASIC witnesses also didn't measure up. The judge rejected chunks of evidence given by Wei Fisher, the translator who claimed she was sacked after MCC said it was pulling out of the deal.

After deliberating for several months, Gilmour handed down his decision two days before Christmas 2009. It was a monumental victory for Forrest. Over the course of 226 pages, the judge found Forrest had not put a foot wrong in the way he described the agreements to the market. According to Gilmour, Forrest honestly believed at all times that he had binding contracts. He added that the Chinese parties had approved the terms of Fortescue's media releases before they were announced to the market.

Gilmour also found the *AFR* article was engineered by He Lianzhong of the NDRC as a "blunt commercial tactic in an attempt to wrest majority control of the project from FMG". Few observers were surprised by this particular finding. China's top business magazine, *Caijing*, had already published an article which implied strongly that MCC made the allegation against Fortescue in retaliation for Forrest's decision to reject China's overtures about equity in the project. According to the *Caijing* article, China's aim was to drive down the Fortescue share price to make it easier for a state-owned company to buy it.

Gilmour said he did not need to reach a view on whether the framework agreements were, in fact, legally binding, as claimed by Fortescue. "I have concluded that FMG's and Forrest's opinion, which underpinned the disclosures, in each case was honestly and reasonably held at the times of the disclosures and thereafter," he said. "This is sufficient for present purposes." A crucial part of the judge's reasoning was his finding that lawyer Peter Huston had been available to advise Fortescue on the nature of the contracts.

Forrest, who was on holiday with his family when the verdict came down, said he was relieved the "distraction" was over. "I'd like to take this opportunity to thank God, the Australian judicial system and my family and friends for their unswerving support throughout the proceedings," he said.

But ASIC soon appealed the decision and the case went before the full bench of the Federal Court for several days of hearings in November 2010. The panel of three judges comprised the chief justice, Patrick Keane, and Arthur Emmett and Ray Finkelstein. On this occasion, Young told the bench that their colleague, Justice Gilmour, had gone "entirely off the rails" by focusing on whether Fortescue and Forrest honestly held the opinion that the agreements were binding when they were clearly not. "We say the question is how would the announcements be understood by a reasonable member of the public," he said. Young also questioned Gilmour's key finding that Fortescue had received legal advice on the statements before they were released.

The result this time was very different. All three judges found against Forrest and Fortescue, overturning Gilmour's earlier ruling. They found the contracts were not binding and Forrest's statements were indeed misleading. They also found Gilmour was wrong to have characterised the announcements as statements of opinion that were honestly held. In his judgment, Keane noted that Forrest had been unable to show any steps he took to ensure that the framework agreements were binding. There was no evidence that lawyer Peter Huston was consulted, he said, and Forrest could therefore not claim the business judgment rule as a defence.

Yet despite finding in ASIC's favour, both Keane and Emmett questioned why the regulator had pursued the case so vigorously when it had not presented evidence to show that anyone had lost money from the misleading statements. "This circumstance may be said to raise a question as to whether the prosecution of this case by ASIC was a game worth the candle," Keane said. Finkelstein, however, could understand why ASIC had brought the case. He was the only judge, in fact, who challenged Forrest's claim that the case was absurd because shareholders had all made money. Finkelstein pointed out what should have been obvious: that anyone who bought shares on the strength of Fortescue's announcements but sold them before they returned to the purchase price many months later had lost money. "More likely than not, many traders lost money and substantial sums of money at that," he said. Finkelstein said that if Fortescue's arguments were accepted, the continuous disclosure laws could be sidestepped by any company whose share price happened to climb after investors discovered that they had been misled. "That is not what Parliament had in mind," he said.

Forrest reacted angrily to the verdict, vowing to remain at the helm of Fortescue and describing ASIC as "mean and vengeful". He promised to appeal the case to the High Court and said the ruling, if upheld, would set a dangerous precedent for the nation. "It's on the record that no shareholder lost money [from Fortescue's actions]," he said. Forrest said he had received "hundreds" of messages of support, both from people he didn't know as well as from leading business people and even members of the judiciary. "Let's just say people are looking at this decision and saying, 'Well, what's really good for Australia?'" he said.

For the first time, however, Forrest admitted that he could have better handled the announcements back in 2004. "When you look at Fortescue, which [at the time] was probably myself and a dog … we could have done so much better," he said. "We should have had lawyers and we should have had legal teams and whole departments, but we didn't. We were a start-up company. We had none of those resources and did the best we could, but we certainly believed a deal was a deal." Forrest's point had some validity, but he was stretching the truth in arguing that in 2004 he was working with little help. The Federal Court was told that Fortescue employed fifty-seven people at the time.

Finally, the case came before the nation's top court. At 10.15am on Tuesday 2 October 2012 – eight years after the ASX announcements had been made – five judges of the High Court delivered their verdict in Canberra. The result was a triumph for Forrest, who was still dusting himself off after the sudden collapse in the iron ore price a few weeks earlier that had threatened to ruin Fortescue. Forrest, who by that time had stood down as Fortescue chief executive, was now free to remain as chairman without the threat of a banning order. He could also continue to serve on other boards, including those of his various charities.

Yet again, the mercurial entrepreneur had gone to the brink of devastation and managed to survive. He had beaten the system that tried to destroy him and had the last laugh over the establishment, which had despised him for so long.

Forrest was holidaying with his family in the bush when a call came through on his mobile phone from Peter Huston. The Fortescue lawyer was just as nervous as Forrest about the outcome; he had advised his boss early in the legal battle that he should fight the charges because he had a good chance of winning. Huston recalled the joyous phone call: "I said, 'We won, Andrew.' He said, 'What? Nicola, come and listen, come and listen!' I said, 'We won, Andrew, we've won. Won everything, everything. Five nil, everything.'" He added: "I knew how much it meant [to Forrest]. There's not been a more important day in my life as a lawyer than that one."

The High Court's reasons for finding in favour of Forrest were very different to those used by Gilmour three years earlier. In fact, its logic in dismissing ASIC's case was a source of deep puzzlement to many observers. The court found that those investors reading Fortescue's statements in 2004 had not been misled because they were savvy enough to have realised that the parties had entered into agreements that would not ever be enforceable in an Australian court. The agreements, which were made with Chinese state-owned companies and signed in Beijing, were more likely governed by Chinese law, the judges found. This was an argument even Fortescue's battery of highly remunerated lawyers hadn't pursued. The court also ruled that Fortescue's statements were not misleading because they were agreements between the parties about what they intended to do in the future.

One of the court's more conservative judges, Dyson Heydon, pointed out that Fortescue's statements were made to a sophisticated audience of investors who knew the perils of doing business with China and understood the risks inherent in promises made by WA mining companies:

> Fortescue's remarks were not directed to the public as a whole. They were directed to a section of the public. It comprised superannuation funds, other large institutions, other wealthy investors, stock brokers and other financial advisers, specialised financial journalists, as well as smaller investors reliant on advice. This was not a naïve audience. It was not an audience in whom the adjectives "Western Australian", "mining" and "Chinese" would excite a sudden certainty about the imminent creation of wealth beyond the dreams of avarice. It was an audience conscious of the difficulties of creating infrastructure for mining projects in the harsh conditions of Western Australia. It was an audience conscious of their vast expense. It was an audience conscious of the problems of doing so in cooperation with a Chinese group described in the ASX announcement as China's largest construction group.

Melbourne fund manager John Robertson described elements of the High Court's judgment as "absurd" and said it placed a fresh burden on investors to second-guess the veracity of company announcements. "They are saying that everyone who is contemplating an investment in the Australian equity market has the skill to differentiate between a measured statement and a company's intentions – that is patently absurd," he said. "If that is now the law, it has been radically rewritten."

A leading corporate lawyer, Bob Baxt, said any suggestion that investors should have realised the contracts were not legally binding did not take into account that continuous disclosure laws apply to all investors, not only those with sophisticated knowledge. "If remarks are made to the ASX they are made to the investing public as a whole," he said. But John Keeves, a partner at Johnson Winter & Slattery, said the High Court had interpreted the market announcement in a realistic way. "The majority decision shows some keen commercial insight from the High Court as to how the announcements would have been understood by investors," he said. "The possible alternative of an overly legalistic approach to disclosure by listed companies could have had disastrous implications for Australia's financial markets."

Whether ASIC's almost eight-year legal pursuit of Forrest was worth the anguish and huge expense – as much as $30 million in legal costs – is debatable. The corporate regulator emerged bloodied and bruised from the encounter, particularly because the High Court in its ruling found ASIC had pleaded its case atrociously – a finding that also reflected poorly on Mallesons, the law firm that ran its prosecution. After the judgment, ASIC warned that its entire disclosure regime would have to be reviewed. "Compliance with continuous disclosure goes to the heart of ASIC's strategic priority of fair and efficient financial markets. We will now assess what impact the High Court's decision has on disclosure requirements," ASIC said.

Living under a serious legal cloud for so many years took a personal toll on Forrest and his family. According to family friends, his youngest daughter, Sophia, was bullied at school after the full bench of the Federal Court found against him in 2011. Forrest and his lawyers came to believe that ASIC had pursued the case with a fanaticism that did not reflect well on its status as a model litigant. "Everyone saw straight through those Chinese playing hard ball in order to get control of an Australian project," Forrest said after the judgment. "But only ASIC saw that I was a high-profile person, someone which they could really pin a scalp to the top of the door if they were successful. Only ASIC wanted to believe that somehow this was some grand conspiracy."

A day after the High Court verdict, Forrest picked up the phone and called ASIC's acting chairman, Belinda Gibson, who was doubtless surprised to find herself chatting to the man she had tried to bring down for the previous seven and a half years. In typically brazen style, Forrest offered to help ASIC learn from its mistakes and to help "create an environment where globally people want to invest and people want to work with ASIC". He told the *Sydney Morning Herald* that he was now "just another Australian concerned that there be an efficient and effective regulator". He would make himself available for workshops to analyse where ASIC went wrong. A cynic might argue that Forrest's gesture was part of a deliberate strategy of keeping his friends close but his enemies closer. Or perhaps he was simply rubbing ASIC's nose in the fact that he was the winner. It should come as no surprise that ASIC has never taken up the offer.

Two weeks later, in mid October, Forrest threw a party at his Cottesloe home to celebrate the victory. Among the guests were Forrest's most trusted legal personnel: his barrister, Allan Myers, and his long-time solicitor, Gadens partner James Scovell. Others present included long-time Fortescue loyalists Peter Huston, Graeme Rowley, Chris Catlow, Mark Thomas and Russell Scrimshaw. Fortescue chief executive Nev Power and chief financial officer Stephen Pearce were also there, representing the miner's new breed of executives.

The biggest shock of the night was the presence of the recently sacked Julian Tapp, Rod Campbell and Ann Marie Lowry, who had all paid the price at Fortescue for the collapsing iron ore price in September. It was Forrest's attempt at an olive branch. The three former executives might have had good reason to stay away, but they put aside their anger to congratulate their old boss on his triumph.

During the party, Forrest's two teenage daughters, Grace and Sophia, spoke about the impact of the legal battle on the family. His son, Sydney, entertained the guests by playing the saxophone. The lawyers regaled guests with stories from the courtroom and there were plenty of laughs. Forrest concluded by thanking everyone there for supporting him, but he saved his biggest praise for God, who he said had helped him win the case. The many unbelievers in the audience shifted uncomfortably at the suggestion Twiggy had the Lord on his side.

In the end, the victory over ASIC was comprehensive. But it still failed to eliminate the doubts many people had about Forrest's truthfulness – doubts that have been reinforced over the years by a series of separate court rulings. One old friend describes the ASIC case as a product of Forrest's inability to tell a story without exaggerating. "Andrew gilds the lily, but in the whole scheme of Andrew gilding the lily that [the ASIC case] was a minor thing," he says. A former Fortescue executive who worked closely with Forrest for several years says: "Seventy per cent of what he says is correct, the other 30 per cent is froth."

Forrest's old stockbroking colleague Jeff Braysich has a similar explanation: "Once he says something the second time, he truly believes it, and that's why he is such a good salesman. The second time he says it, he thinks, 'I've heard that before,' but he forgets it was he himself who said it." Former colleague David Mendelawitz believes Forrest's ability to embellish a story has been integral to his success. "He's such a good salesman – you just don't know what's fact and fiction," he says.

Rodney Adler, who went to jail for corporate fraud, is one old friend prepared to assert that Forrest is an honest person. "I know what a lot of people have said about Andrew – I've met a lot of them and I've had to mediate arguments between people and Andrew over the years," Adler says. "To me, Andrew has never once broken his word. To me, you could not get a more honest person with more integrity." Adler concedes, however, that Forrest would probably do things differently if given the chance again. "When you start with nothing and you build one of the greatest mining empires in the world, are there things you would do differently in hindsight if you had more time, more money and more experience?" Adler says. "Yes, there are. But sometimes you've got to do things to survive that you would have preferred not to do."

# 12.

## TAXING TIMES

*I'm very happy to pay tax. In fact I'm honoured to be one of Australia's major taxpayers.*
—ANDREW FORREST

Being crowned Australia's wealthiest man in 2008 did wonders for Forrest's public profile, even though he insisted he'd prefer to remain anonymous. "I really don't enjoy the spotlight," he said after topping the *BRW* Rich List with a fortune approaching $10 billion. "And I particularly don't enjoy the attention which the accident of my wealth creates. I may have become Australia's richest man, but to me that is an incredibly uninteresting title." Without a hint of irony, the billionaire who had accumulated the nation's fastest fortune thanks to his gift of the gab declared he was reluctant to talk about his achievement. "I don't want to look as if I am out there trying to beat the drum because that is not what I am trying to do."

But Forrest was soon to discover one of the great benefits of his newfound wealth and success: from the prime minister down, people tended to listen to what he had to say. When he needed to "beat the drum", an art form in which he was particularly skilled, the national media responded by lapping up every word uttered by this maverick entrepreneur who had emerged from the west. It wasn't until 2010, however, that Forrest abandoned his efforts to keep a low profile – assuming he had ever tried in the first place –

and thrust himself into the middle of the most high-stakes clash in recent Australian political and corporate history.

On 2 May 2010, prime minister Kevin Rudd and his treasurer, Wayne Swan, strode into a heaving media conference in Canberra to announce the creation of the resource super-profits tax (RSPT). From now on, they declared, mining companies reaping astronomical earnings from the greatest resources boom in history would have their "super profits" – defined as anything above the government bond rate – taxed at 40 per cent. The miners would still also pay the normal rate of company tax on their profits, although they would have the royalties they paid to the states refunded under the new regime.

To many, slapping a new tax on the industry seemed reasonable: the miners were growing rich beyond their dreams, thanks to the soaring value of the finite resources they were digging up and shipping out. By 2010, the value of Australia's iron ore exports alone had reached an astonishing $50 billion – a tenfold increase in just a decade. In theory, a resource rent tax recognises that a mining company's profits are partly the result of its own efforts and partly the result of the value of the natural resource it has a licence to exploit. Even the big miners had conceded before 2010 that they could pay more tax; their peak body, the Minerals Council of Australia, had previously backed the concept of a resource rent tax as a replacement for state mining royalties.

Rudd and Swan promised the RSPT would raise $12 billion in its first two years from July 2012 – the sort of revenue bonanza that would leave any political leader licking their lips. The money would be used to increase the superannuation guaranteed rate for all workers from 9 per cent to 12 per cent of their salaries, lowering the company tax rate and boosting infrastructure spending in the mining states of Western Australia and Queensland. As treasurer, Swan had allocated all of the funds before he had even collected them.

Rudd and Swan also endorsed Treasury secretary Ken Henry's hypothesis that a resource rent tax would help Australia avoid the phenomenon of "Dutch disease", in which a strong resources sector drives up the local currency, leading to a decline in manufacturing and other trade-exposed industries. Australia was awash at the time with anecdotes about the negative impact of the "two-speed economy" as Western Australia and Queensland powered ahead of the other states. Rudd and Swan had another reason to put their faith in Henry: the powerful bureaucrat had proved invaluable in advising them in 2009 to stimulate the economy during the global financial crisis, a move which had helped Australia to emerge from the turmoil relatively unscathed.

When the mining companies began to look more closely at the Labor government's RSPT proposal, they were shocked by what they saw. For a start, they claimed, the headline rate of 40 per cent was simply too high and would kill marginal projects and deter future investment. They also believed it was unfair that the tax would be applied retrospectively on projects in which they had invested decades earlier. And there was another deep flaw. The keystone of the RSPT was that the government would take a 40 per cent equity interest in all mine projects, and any so-called super profits made on that 40 per cent would flow to Canberra. But even more radically, the government was also proposing to refund 40 per cent of any losses that a project might suffer.

To Forrest, this amounted to nationalisation of the mining industry by stealth. He suggested the guarantee to refund losses was worthless, as banks would never finance projects based on the promise that 40 per cent of losses might one day be reclaimed through the tax system. Worse, he argued, it could lead to "mums and dads" forking out billions of dollars to bail out unprofitable projects. Forrest believed the tax had been designed by officials in Canberra with little knowledge of project financing or the realities of the mining industry. The Treasury boffins had come up with a tax that was theoretically elegant but practically useless.

It took only a few hours before the three heavyweights of the Australian resources industry – BHP Billiton, Rio Tinto and Swiss miner Xstrata – decided the RSPT had to be killed. Bizarrely, none of the multinationals had been consulted over the detail of the proposed new tax even though they would be paying the bulk of it. Worse, they had been led to believe by Swan in the days before the 2 May announcement that the RSPT had not been finalised and no revenue had been booked in the federal budget to be handed down on 10 May. Both of these assurances turned out to be false, which left the miners feeling betrayed. Almost from day one, it was obvious – even to senior members of the government – that the RSPT could not survive.

In her 2012 book *Tales from the Political Trenches*, former Labor MP and Rudd ally Maxine McKew described Swan's "flawed execution" of the RSPT and said the big miners "felt blindsided by an uncompromising treasurer". Rudd was another who felt betrayed by Swan, who had led him to believe the miners were onside and that the premiers of the two big mining states – Colin Barnett in Western Australia and Anna Bligh in Queensland – had agreed to the proposals. "Swan had not delivered and Rudd had come to believe he had been sold a pup," McKew wrote.

Swan had also overseen the design of the new tax without consulting any of his ministerial colleagues, several of whom would have been able to point out its flaws. One of them, resources minister Martin Ferguson, was so irate when told about the RPST three days before it was announced that he considered resigning. Ferguson had spent the previous few months promising the industry that it would be fully consulted over the detail of any tax changes. He felt as if he had been hung out to dry.

As McKew wrote, Swan should have spoken to Ferguson as well as Gary Gray, who had close contacts in the resources sector in Western Australia, and Craig Emerson, an economist who wrote his doctoral thesis on the Hawke government's successful introduction of a resource rent tax on petroleum projects in the 1980s. "The combined expertise of Gray, Ferguson and Emerson could have been harnessed earlier by Swan to work with industry, get the detail right and arrive at a sensible, middle-ground position – one that industry could live with, and one that achieved the primary purpose of an appropriate return for the community from the right to mine the country's non-renewable resources," McKew wrote.

In the days after 2 May, BHP and Rio went on the warpath, claiming that the RSPT represented the biggest political risk they faced anywhere in the world. BHP chief executive Marius Kloppers, an idiosyncratic South African who was based in Melbourne, estimated BHP's effective tax rate would climb from 43 per cent to 57 per cent under the RSPT, compared to a rate of 23 per cent in Canada and between 27 and 38 per cent in Brazil. Under this sort of tax regime, Kloppers argued, BHP was unlikely to invest much in Australia in the future. To some, this sounded like scaremongering on a grand scale. To others, the companies were expressing legitimate concerns about a tax that appeared overly punitive and badly designed.

When Forrest and his bookish chief financial officer, Stephen Pearce, began digesting the complex detail of the RSPT, they soon realised the impost posed an even bigger threat to Fortescue than it did to their bigger rivals. Forrest had been tipped off by trade union leaders several weeks earlier that the government was planning a new resources tax. The industry had also been alerted on 24 April by a front-page article about the planned tax by the *Australian*'s political editor, Dennis Shanahan. Forrest sought assurances from

Swan in the days before 2 May. "I rang the treasurer three times," Forrest said, "and each time he assured me the tax would be benign, that it would be no problem for Fortescue. Frankly, I was not reassured."

Fortescue was acutely exposed to the RSPT because it was built almost entirely on debt and had yet to repay its billions of dollars in loans. The company's business model was vastly different to those of BHP and Rio, which could fund their fresh investments from their cash flows. The money Fortescue needed to repay its bankers would effectively be eaten up in RSPT payments. Pearce set about modelling the impact of the RSPT and presented his findings to Forrest, who was horrified. "It showed we would have a negative cash flow after paying interest," Forrest said. "It would lead Fortescue to a default situation. Our company would not exist under the RSPT. My understanding is that the requirement in the US capital markets is that once you face difficulty in meeting any interest payment you have an obligation to inform bondholders and that triggers a potential default event."

For Forrest, the RSPT represented the single biggest threat to the profitability of the company he had founded seven years earlier. He began describing it as "the worst economic policy in Australian history". In the days after 2 May, Fortescue's share price was hit harder than that of any other miner as about $20 billion was wiped off the value of the Australian sharemarket. Analysts at JP Morgan estimated that Fortescue's net present value – an assessment of long-term profitability – would suffer a hit of 20 per cent under the RSPT, compared to only 10 per cent for BHP and Rio. A research note published by Macquarie Bank downgraded the value of Fortescue's Solomon and Chichester expansion projects by 20 per cent and 35 per cent respectively.

Forrest's friendship with Rudd was the first casualty of the mining tax uproar. The pair had got on well since meeting years earlier at the Perth dinner organised by Brian Burke to introduce Rudd to local business leaders as a future ALP leader. In 2008, when Rudd was prime minister, the friendship was cemented when he backed Forrest's plan to create 50,000 jobs for indigenous Australians through the Australian Employment Covenant. That year, the pair also caught up at the Olympic Games in Beijing and at Rudd's 2020 Summit in Canberra. Forrest, a natural political conservative, had been close to John Howard, but he had become even closer to Howard's Labor successor.

By sheer coincidence, Rudd had long been scheduled to meet a group of WA's senior business leaders, including Forrest, at a dinner in Perth on Tuesday 4 May – two days after the RSPT was unveiled in Canberra. What was originally planned as a polite conversation turned into an ambush. Forrest led the charge, presenting Rudd with a set of red boxing gloves with the words "Fair suck of the sauce bottle" scrawled across them. It was an attempt at humour – Rudd had been ridiculed for using the phrase in public statements – but the gesture was also deadly serious. As Forrest and the thirty other guests dined on Red Emperor fillets and quaffed bottles of 2004 Voyager Estate cabernet sauvignon, Rudd was forced to listen for hours to the miners telling him bluntly that the RSPT would kill their industry and cost jobs. But Rudd was unmoved, telling the miners that some of the finer details could be tweaked during consultations, but the 40 per cent tax rate and other key elements of the RSPT were non-negotiable.

Forrest decided to go on the attack. Sitting at his desk in front of the national flag, he appeared on the ABC's *Inside Business* program to claim financiers were "fleeing" Australia and that Fortescue had shelved its planned expansion projects in the Pilbara, putting 30,000 jobs at risk. These were claims that were difficult to disprove, and they made good headlines. Forrest also claimed he had been reluctantly drafted to the debate. "I'm out there employing thousands of people, building billions of dollars' worth of projects as we speak, loving my job," he said. "I don't want anything to do with this political process. I've always thought it was well managed by people who understood, but unfortunately these people do not understand what it takes to build industries and we are suffering as a result."

One week after the RSPT announcement, the Minerals Council of Australia launched an astonishingly effective advertising campaign against the tax. Rudd and Swan attempted to retaliate with some government-funded ads, outlining why Australia needed an RSPT, but they were no match for the industry's $22-million prime-time blitz. Forrest, who had refused to join the MCA because it was dominated by his enemies at BHP and Rio, formed a loose alliance with smaller miners and explorers in Perth to lobby against the tax. This campaign was run through the Perth-based Association of Mining and Exploration Companies, which had hundreds of members. But Forrest emerged as the self-appointed flag-bearer of the "small" miners' revolt against the tax, even though Fortescue had grown to become a major iron ore exporter. He made himself available for interview after interview, generating free media coverage for the cause. He even seemed to be enjoying it.

Rudd responded to Forrest's daily media appearances by saying people should take the billionaire's comments with a "grain of salt". By early June, Rudd had stopped returning Forrest's phone calls, accusing him of campaigning against the RSPT to avoid "paying a dollar more in tax". Forrest insisted he was happy to pay more tax as long as the system was fair. "I'm very happy to pay tax, in fact I'm honoured to be one of Australia's major taxpayers," he said. How Forrest arrived at this conclusion was puzzling, because at that time he had never actually signed a corporate tax cheque. This was because Fortescue had never made a taxable profit, thanks largely to the legitimate deductions for financing and exploration costs available to start-up miners. Forrest did not send his first corporate tax cheque (which was for $140 million) to Canberra until late 2011.

On 9 June, Forrest shared the stage with his *bête noire*, Gina Rinehart, at what was to become one of the seminal events of the battle: the lunchtime anti-tax rally in Perth attended by 2000 people. The rally was held in a park behind the Hyatt hotel, where Rudd was about to deliver a speech. It also happened to be next door to Fortescue's head office, ensuring that most of the company's staff were among the placard-wielding audience. Never before had a protest against injustice been led by two of the richest people in the land. But for once Forrest wasn't the centre of attention, as the normally reclusive Rinehart – who had donned her finest pearl necklace for the occasion – climbed on the back of a flatbed truck and screamed herself hoarse, leading the compliant crowd in a repetitive chant of, "Axe the tax!"

Clad in the high-visibility overalls he normally wore only in the Pilbara, Forrest at least looked like a real miner. He skilfully weaved his own rags-to-riches story as he worked the crowd into a frenzy while attacking Rudd, who he suggested was "turning communist" by imposing the RSPT. "Fellow miners, fellow Australians … this day is about your opportunity to change Australia's history from where Kevin Rudd would take it – a largely socialist distribution of capital over creation of value," he said. He also revealed that he had finally secured a meeting with Rudd the next day to discuss the tax.

Forrest then strolled over to the Hyatt to attend the lunch at which Rudd was speaking. His anger had barely had time to subside before the embattled prime minister began invoking the spirit of Sir John Forrest to justify his own plan to use the RSPT to build infrastructure for future generations. "John Forrest was determined to ensure the great mining boom of the 1890s built enduring foundations for WA's future," Rudd told his lunchtime audience. "Forrest supported mining. He supported it strongly but was often opposed by gold miners of his time. As his biographer Frank Crowley has noted, Forrest wanted to make the prosperity of the present pay for the hoped-for success of the future. This did not endear him to the gold miners of the time … some things never change." Rudd was brazenly comparing himself to the visionary Sir John, the great-great-uncle whose memory Forrest had long revered. It didn't appear to be a good omen for the next day's peace talks.

Few realised at the time, however, that a desperate Rudd had already decided Forrest needed to be brought onside if there was to be any hope of negotiating a truce in the mining tax war. The 10 June meeting in Perth was a turning point. Rudd invited Forrest to enter confidential talks with his office to seek a solution. After the meeting, Forrest's tone was suddenly conciliatory. "I have a prime minister who is listening and who has entered into a process," Forrest said. "From my perspective, we have much more wood to chop. He was open, convivial and the discussions were detailed. He's open to a process, and let's give that a chance."

Ten days later, Forrest was whisked into The Lodge alongside Stephen Pearce to attempt to broker a solution to the crisis, which by then was beginning to threaten Rudd's increasingly chaotic prime ministership. Rudd's aim was to split the mining industry by finding an outcome that would appease Fortescue but ensure the multinationals still paid the bulk of the new tax. All of a sudden, Forrest was being given the chance to devise a plan that would ensure his own financial future, damage his enemies and keep Rudd in office. Nine years after his reputation had been shot to pieces and he'd been banished in the wake of the Anaconda debacle, Forrest was now operating at the apex of national power.

Forrest's original push was to win Rudd's endorsement for his idea that Fortescue should be able to pay its massive interest bill before the RSPT kicked in. But a series of emails, later released under freedom of information laws, showed that Treasury officials were violently opposed to this concept because they feared it would skew investment decisions in favour of debt rather than equity. "Collectively, the proposals would ensure that FMG would pay very little, if any, RSPT in excess of existing royalties," wrote Scott Bartley, Treasury's principal adviser in the resource tax unit. "No other 'normal' business gets the capital repaid before a tax kicks in – still that is what FMG appears to be asking for."

Rudd refused to back down on the claim for interest deductibility or the 40 per cent headline rate, fearing he would be crucified by the electorate for giving away too much to the miners. But he was keen to listen to a novel plan Forrest had hastily devised, under which mining companies would be given tax breaks for their infrastructure spending. Crucially, under Forrest's plan, such deductions would be available only to miners such as Fortescue that built open-access infrastructure, rather than BHP and Rio, which had barred anyone else from using their railway lines.

The restructured RSPT, as envisaged by Forrest, would also have benefited Fortescue over its rivals by favouring miners that had recently built their infrastructure. In a briefing note sent to the Fortescue board after Rudd had agreed to the plan, Pearce summarised the impact of the proposal: "In essence, while a headline rate of 40 per cent is maintained, with the concessions negotiated, an effective headline rate of 20 per cent with full royalty rebate has been achieved."

Forrest insisted he would have sought the agreement of BHP and Rio before signing any deal with Rudd. "My position was that Fortescue would not sign any separate agreement with the government," he said. "We wanted a solution for the industry. It was my intention to take the principles agreed with Rudd to the industry." Yet Forrest must have known that BHP and Rio would never have supported the deal.

Back in Perth, Forrest's allies in the war against the RSPT were gobsmacked when they discovered that the magnate had been negotiating a backroom deal with Rudd. Atlas Iron's David Flanagan said he knew nothing of the talks and was not convinced the Forrest–Rudd pact would have benefited his fledgling iron ore company. "I am concerned that all of the companies were not at the table and have not been briefed," he said. "It's important to work together." Another close ally, BC Iron's Mike Young, said the proposals were not necessarily good for everyone. "What is good for FMG is not going to be good for BC Iron," he said. A senior Labor government official summarises Forrest's motivations in pursuing the deal with Rudd: "Twiggy was trying to do a deal that advantaged his own company, but it wouldn't have advantaged any of the others."

The deal with Forrest would not have stopped the advertising campaign against the government, which was driven by BHP, Rio and Xstrata. But it would probably have relieved pressure on Rudd's leadership for some time. A government official who was familiar with the strategy describes Rudd's thinking in pursuing the Forrest plan: "We all thought Twiggy was a cowboy. But when it became absolutely clear there was no deal with the others, and Twiggy was rattling around Canberra, we thought, 'We have an opportunity here.' Because of the nature of Twiggy's business we could write a rule that gives him a big benefit that is not available to the big guys, and which preserves the integrity of the tax. We couldn't have done a deal with BHP, Rio and Xstrata because they are the ones paying all the tax. They're the ones who have the massive cash cow mines where they built the infrastructure ages ago and have just been milking massive windfall profits out of high commodity prices.

Making them happy would have killed the tax. For us, it would have been a good story – the newer miners would get a benefit and we'd still get the revenue because most of the revenue would come from BHP, Rio and Xstrata. Simultaneously, because Twiggy was such a larger-than-life figure and everyone identified him with mining, it was a great win. We'd be able to shake hands with Twiggy, make a change to the policy which was even more favourable for investment and the development of new mines, and have someone who Australians would have thought, 'If he is shaking hands with Rudd, it must be okay for the miners.'"

That a multibillion-dollar tax policy was being cooked up in this manner speaks volumes about the frenzied dying days of the first Rudd government. Still, the deal came painstakingly close to being done. Forrest and Pearce first met Rudd and his youthful economics adviser, Andrew Charlton, at The Lodge to discuss the plan on the evening of Sunday 20 June. After a glass of wine, the talks lasted until the early hours. But the full details still needed to be ironed out, so Forrest and Pearce were invited back to conclude the negotiations the next evening. This was a highly chaotic period in the halls of power in Canberra. Labor MPs had already begun secretly counting the numbers for a coup against Rudd that would install his deputy, Julia Gillard, as prime minister later that week.

On the next day – Monday 21 June – a huge Chinese delegation led by then vice-president, and future president, Xi Jinping was in Canberra for talks with Rudd and business leaders, including Forrest. The meeting between Rudd and Forrest planned for that night did not occur because Rudd spent several hours longer than scheduled chatting in Mandarin to Xi. To the frustration of everyone except Rudd, Forrest was asked to come back again the next night to finalise the deal.

But early the next morning, Forrest received the news that the body of his friend Craig Oliver had been recovered from the wreckage of a plane crash in Africa that killed the entire board of Perth-based iron ore miner Sundance Resources. Oliver had lived around the corner from Forrest in Cottesloe and their daughters attended school together. Forrest flew back to Perth immediately on the Fortescue corporate jet, with the aim of returning to Canberra the following day to consummate the deal. A press conference had been tentatively scheduled for Friday so Rudd and Forrest could front the cameras and hail their breakthrough agreement.

But just as Forrest was preparing to board the plane for Canberra on the Wednesday evening, word began filtering through that Gillard had challenged Rudd. Forrest stayed in Perth, glued to the television. The next day, Gillard became prime minister, citing the RSPT fiasco as a major reason for her decision to seize control of a government that had "lost its way". Despite his culpability for much of the RSPT debacle in the first place, Swan retained his job as treasurer and was even promoted to deputy prime minister.

Forrest had come agonisingly close to striking a deal with Rudd, but a series of unrelated events had conspired against him. "If Rudd hadn't talked to bloody Xi Jinping for four hours on the first night, and if the Sundance plane hadn't gone down on the second night, and if Gillard hadn't moved on Rudd the third night, the deal would have been signed," says a person familiar with the talks.

In her first press conference as prime minister, Gillard promised that the government would suspend its advertising campaign if the big miners did the same. Amazingly, a truce had been declared within minutes of the press conference. What the new prime minister didn't say was that the ceasefire – at least in the opinion of many of the key players – had been pre-arranged with the miners before the coup.

In the days before 24 June, as Forrest attempted to reach an agreement with Rudd, the big miners were simultaneously talking to Wayne Swan and Martin Ferguson, with the aim of reaching a deal of their own. Gillard was kept informed of the progress of these talks, although she was not directly involved. Under the planned agreement, BHP, Rio and Xstrata would have been able to assess their decades-old mining investments at their current market value rather than their much lower book value, giving them huge financial relief and addressing their concerns over the retrospective nature of the tax.

To what extent Rudd knew of these negotiations – and whether he would ultimately have agreed to any sort of deal with the majors – remains a matter of conjecture. But several people familiar with the situation are adamant that Gillard and Swan knew they had the big miners onside when they removed Rudd on Thursday 24 June. "[Then ALP national secretary] Karl Bitar had cooked up the idea of the truce," says one official who was close to the action. "They had pre-agreed the idea of a truce, that Gillard would stand up and say, 'I will withdraw the government ads and I ask the mining companies to withdraw their ads.' I don't think that was a risk Gillard took – I think it was a pre-cooked deal."

Within a few days, Gillard had negotiated a brand new tax with BHP, Rio and Xstrata that would be known as the minerals resource rent tax (MRRT), with an effective headline rate of only 22.5 per cent and generous deductions for existing investments. It would apply only to iron ore and coal, the nation's two biggest export earners. The MRRT was a massively watered-down version of the RSPT. Despite this, Gillard hailed the deal as a "breakthrough" that proved she was busy fixing the mess Rudd had created. The government's original forecast of $12 billion in revenue from the RSPT had been cut to $10.5 billion under the MRRT. It beggared belief, however, that the huge concessions Gillard and Swan had given to the big miners in brokering the peace would cost the government only $1.5 billion over two years.

Forrest, inevitably, was irate at being excluded from the post-coup negotiations on the MRRT. His influence in the corridors of power had seemingly vanished overnight, and he was especially aggrieved that BHP and Rio had been invited into the prime minister's office to seal a deal that appeared to suit them at the expense of smaller miners. Forrest claimed the big miners would end up paying very little MRRT because of the generous depreciation allowances they had been given. Swan dismissed this, saying BHP, Rio and Xstrata would pay 90 per cent of the new tax. But mining analysts at leading investment banks who crunched the numbers were also questioning how the government would ever raise $10.5 billion from what seemed a relatively benign impost.

Like others, Forrest believed the multinationals had decided before 24 June that removing Rudd as leader and installing Gillard was the only way to get what they wanted. "We didn't realise that BHP and Rio had gone behind our backs to do another deal," Forrest said. "When we were negotiating our solution with Rudd, I was calling BHP and Rio but they didn't call us back. As soon as Rudd was gone, BHP and Rio didn't need us, an Aussie battler, anymore. They were doing their own deal with Gillard." Gillard and Swan later rejected Forrest's claims that BHP and Rio backed the coup as "a conspiracy theory" and "a rehashing of old nonsense".

If Gillard and Swan thought that was the end of the mining tax fracas, they hadn't counted on Forrest's absolute hatred of losing. While BHP, Rio and Xstrata suddenly fell silent, content with the bland version of the tax they had secured, Forrest refused to lie down. He threatened to challenge the MRRT in the High Court and accused the big miners of negotiating a deal that suited themselves at the expense of companies that were reliant on debt financing to develop new projects. Forrest claimed Fortescue would "carry the burden" of the MRRT, which he labelled "economic vandalism".

Gillard called an election for 21 August and Opposition leader Tony Abbott promised he would repeal the tax if he won office. Forrest, while still trying to run a major listed company, embarked on a national tour of mining communities in Labor-held marginal seats, in an attempt to drum up opposition to the MRRT. He was waging a one-man campaign against the mining tax. "It was a really gruelling time for Andrew – Nicola was concerned how much work he was doing," says a friend.

Forrest insisted all along that he was apolitical and he refused to recommend a vote for the Abbott-led Liberal Party at the election. But he did work very closely with numerous Liberal MPs in fighting the tax. He also put more than $1 million of Fortescue money into setting up a foundation called Keep Australia Afloat, which was run by Liberal Party member Bill Mitchell and aimed to generate broad community opposition to the tax. Colourful Sydney business identity John Singleton was closely involved with the group, having created a multimillion-dollar advertising campaign against the RSPT that was pulled at the last minute when Gillard came to power. Singleton was a long-time friend of Gina Rinehart, who, like Forrest, had put about $1 million into Keep Australia Afloat.

For a few days after the 21 August federal election, it appeared as if Forrest's bid to kill the mining tax had been a success. It was a cliffhanger result and Abbott appeared to be in the best position to become prime minister. A day after the election, with the outcome still in the balance, Forrest said Australians had rejected "irresponsible economic management" and that no party had a mandate to introduce the MRRT. But in the days that followed, Gillard managed to win the support of independents Rob Oakeshott, Tony Windsor and Andrew Wilkie to cobble together a minority government.

Forrest still refused to give up. He launched a fresh effort aimed at convincing Oakeshott, Windsor and Wilkie to vote against the MRRT legislation when it was introduced into parliament. But his pre-election claim that Fortescue would "carry the burden" of the new tax had to be discarded when internal modelling showed the company would hardly pay a cent of the MRRT. Ironically, this was because the generous deductions that BHP, Rio and Xstrata had negotiated for themselves would also prove to be of huge benefit to Fortescue.

Undeterred by the fact that he would escape paying the MRRT, Forrest vowed to continue the fight on behalf of smaller miners, who he said were the victims of a tax that would take from the poor and give to the rich. Forrest was now casting himself as the Robin Hood of the mining industry. The MRRT, he claimed, would affect companies still trying to obtain financing to build new mines. These smaller companies would, proportionally at least, take a much bigger hit than the big miners. "My heart doesn't break for Fortescue," he said. "It breaks for all those other companies that want to be the next Fortescue. The impact won't be next year. In the long term, with the MRRT, we will look back on the resources sector and wonder what happened. It will be like a slow-growth cancer. It will slowly push jobs overseas."

Forrest had two other good reasons to persist with the fight: his bitter hatred of both BHP and Wayne Swan. While Forrest and BHP's Kloppers had managed to put aside some of their differences in fighting the RSPT back in May, the personal relationship between the two men deteriorated as it became clear that Kloppers was the one with the greater power to influence public policy.

Swan, emboldened by his election victory, was also growing tired of Forrest's endless media campaign against the mining tax. From the earliest days of the stoush, it was clear that Swan had little respect for Forrest, even telling him during a meeting in 2010 that if he didn't like the RSPT, he and his fellow West Australians should simply secede. Swan mocked Forrest's claims that the

MRRT would fail to raise anything like $10.5 billion in its first two years of operation. He then took the extraordinary step of writing a 4000-word article for the *Monthly* magazine in which he described the anti-tax campaign being waged by Forrest and fellow mining billionaires Gina Rinehart and Clive Palmer as a "threat to democracy".

In Swan's eyes, the magnates were attempting to sabotage public policy for reasons of their own self-interest, and they were pocketing a disproportionate share of the nation's riches. Swan also pointed out the inconvenient fact that Forrest had admitted to paying no corporate tax at Fortescue (even though he had, in fact, recently done so for the first time). "I fear Australia's extraordinary success has never been in more jeopardy than right now because of the rising power of vested interests," Swan wrote. "This poison has infected our politics and is seeping into our economy. Though these vested interests have not yet prevailed, every day their demands get louder." Swan's essay quickly became big news. Never before had a federal treasurer launched a class war against wealthy business people for doing what came naturally: pursuing their own interests to get rich.

Forrest's over-the-top response to the Swan diatribe illustrated how paranoid he had become. Within a few days, Fortescue placed full-page newspaper advertisements across the country, headlined "Swan's attack is unfair, untrue and divisive" and accompanied by a photograph of Forrest with a group of happy Aboriginal employees. The ad, purportedly written by Forrest's friend and Fortescue deputy chairman Herb Elliott, pointed out that the company was hardly a tax evader. In fact, it would pay more than $1 billion in taxes and royalties in 2012, and this was expected to climb to more than $2 billion in 2013.

Elliott didn't hold back. "Andrew Forrest and his team created one of the great business success stories in Australian history," said the advertisement. "He started with nothing and repeatedly put everything he earned at risk in building one of the most important mining operations in the world. Andrew epitomises the spirit of what an Australian can do if given a 'fair go'. For a politician to suggest that he has lost sight of this 'fair go' ethic is baffling to those of us who know him best and work with him daily."

Swan cited the Fortescue ads as further proof of the way in which Forrest was prepared to use his wealth to influence public policy. "The average punter only gets to vote at the election – they don't get to take out full-page ads like Mr Forrest has taken out today," he said. A close ally of Swan's, Sydney investment banker Paul Binsted, was one of the few in the business community to publicly back the treasurer. "I hardly think that Twiggy Forrest, Clive Palmer or Gina Rinehart represent business," Binsted said. "Personally, I see them as being like lottery winners who confuse their good fortune with skill."

As Liberal MP Malcolm Turnbull – a man who knows a thing or two about making money – accurately pointed out in the *Australian Financial Review* several days later, Swan was himself guilty of bowing down to "vested interests" by allowing the big mining companies to kill off the RSPT, a move that cost the country tens of billions of dollars in lost revenue. Swan had surrendered to the miners, Turnbull alleged, because he was desperate to retain power. "It is Swan's job (and the job of all of us in Canberra) to stand up to 'vested interests'," Turnbull wrote. "To weigh their claims and respond in the national interest. But in truth Swan assailed [Forrest, Palmer and Rinehart] not because they asserted their interests, but because they are the most inviting and politically expedient targets in the broader mining industry which so crushingly routed the treasurer and his department over their incompetently designed resources tax in 2010."

The MRRT finally came into force on 1 July 2012. Nine days earlier, Forrest announced he would fund a challenge to the tax in the High Court on the grounds that it was unconstitutional. To many, it looked as though he was fighting a war long after everyone else had packed up and gone home. Few could understand his fixation on killing the MRRT when Fortescue would not be liable to pay it for at least another five years and it seemed most others would also be unaffected. Forrest insisted he was fighting the MRRT because of its inherent "evil". Paraphrasing a quote widely attributed to the eighteenth-century statesman and philosopher Edmund Burke, Forrest proved he would never knowingly understate things. "All it takes for evil to conquer is for good people to do nothing," he declared. "We've decided not to do nothing."

Fortescue's challenge was based on the claim that the MRRT discriminated between the states on taxation grounds and was therefore in breach of section 51 of the Constitution. But in a written submission to the court, the government said the High Court should see the challenge as an example of Forrest attempting to evade paying the MRRT. "In the end, this case should be seen for what it is," the submission said. "The plaintiffs do not like the MRRT Act because they may have to pay more for the ability to harness above-normal profits from carrying on mining in Australia."

When the case came before the High Court in March 2013, Fortescue's barrister, David Jackson QC, had trouble winning support from the panel of judges. Jackson argued that because mining royalties – which vary between the states – could be deducted from a company's MRRT liability, the new tax was being imposed unequally in each state. Some of the judges pointed out during the hearing that companies were already able to claim deductions that reduced what they paid in company tax, and there was no suggestion that these other commonwealth taxes were illegal. In August 2013, just weeks out from the federal election, the High Court confirmed that the challenge had failed and the tax was constitutional. When the judgment was handed down in Canberra, Forrest was in Kalgoorlie at the annual Diggers and Dealers mining conference, where he received an award for services to the industry and was fêted by the horde of hopefuls who attend the bash each year. He insisted he had no regrets about the challenge and denied the case had been a waste of Fortescue shareholder funds. Forrest was putting on a brave face, but his three-year crusade against the MRRT had finally come to an end.

For all his bluster and swagger, however, Forrest emerged from the mining tax war with more credibility than Wayne Swan. That was not difficult. Figures released in the May 2013 federal budget showed that the MRRT – which was introduced to "spread the benefits of the boom" – would raise a pitiful $200 million in its first year of operation. BHP, Rio and Xstrata paid little or no MRRT in the first twelve months, despite all three making billions of dollars in profits from their Australian operations. The $200 million collected was the equivalent of loose change rattling around in the Treasury coffers. Swan had originally budgeted to collect $4 billion in MRRT revenue in 2012–13, but later revised this down to $3 billion and then to $2 billion. He blamed lower commodity prices for the huge shortfall and insisted that revenues would improve over time. But most believed that Swan and Gillard had given away far too much to the big miners in the desperate days around their 2010 leadership coup. Another reason for the failure of the MRRT was Swan's blunder in agreeing to refund mining companies for any increases in state royalties that were introduced after May 2010 – a move that gave the premiers an incentive to hike up their royalty rates, knowing that the commonwealth would pick up the tab.

On a sizzling day in the middle of February 2013, Forrest was in the Pilbara when a call came through on his mobile phone. On the other end of the crackly line was Andrew Wilkie, one of the independents who had put Gillard into power and backed the government's MRRT. Wilkie was calling to apologise to Forrest for ignoring his repeated warnings that the mining tax was a dud.

Wilkie went on ABC radio later that day to demand Swan follow suit and apologise. "I think Andrew Forrest is owed an apology because in the lead-up to the MRRT becoming law, he lobbied very hard the crossbenchers [and] the government, I assume," he said. "And he argued there was a fundamental flaw in the tax, and that was the way the depreciation provisions were set up and they unfairly favoured the big three miners. Now, I must admit at the time I was sorely tempted to go with his line of thinking. Instead I decided to trust the Treasury and the government, but he's been proven completely right." Wilkie, a former Green from Tasmania, was not known to be pro-mining and was hardly in the same ideological camp as Forrest. But he was finally conceding his mistake in supporting what many observers regarded as one of the worst examples of public policy-making in Australia for many years.

Forrest might have failed in his quest to steer government policy in his favour, but the experience was nevertheless salutary. In Forrest's mind, the fight to end the "evil" of the mining tax was no different to any of the other noble causes he champions – like his crusade to end Aboriginal impoverishment or his indefatigable quest to create nation-building wealth and jobs at his Pilbara mines. After Wayne Swan grouped him with Gina Rinehart and Clive Palmer as one of Australia's greedy, self-interested billionaires, Twiggy began privately referring to the trio as "the Good, the Bad and the Ugly", casting himself as the only honourable mining magnate.

He is particularly keen to portray himself as a reluctant meddler in the dirty political process, as someone who stands above the shadowy world of party politics and backroom deals. Yet the innate deal-maker in Forrest had jumped at the chance to negotiate a private mining-tax agreement with Rudd that would have served his interests more than anyone else's. The truth is that politics is a natural fit for Forrest's salesmanship and his impulse to change the world. It's a calling he may one day be unable to resist.

# 13.

## THE INTERVENTION

*The more you know Aboriginal people,
the more you love them.*
—Andrew Forrest

As he worked feverishly to build a mining empire in 2005, Andrew Forrest began to develop a close interest in Yindjibarndi country, a vast swathe of the Pilbara centred on the gorges, waterfalls and palm-fringed rock pools of the spectacular Millstream-Chichester National Park. That year, after a decade-long legal struggle, the Yindjibarndi people finally established non-exclusive native title rights over much of the region. Federal Court judge Robert Nicholson had earlier recognised the Yindjibarndi's "remarkably strong sense of connection" to the land, which stretched back as long as 40,000 years.

Forrest had already acquired tenements to explore for iron ore over about half of the 13,000 square kilometres of the native title area. But in 2005, his geologists at Fortescue hit paydirt when they unearthed a massive deposit of iron ore in the Hamersley Range, in the southernmost part of Yindjibarndi country. The deposit was christened Solomon, after the biblical king who made a fortune from mining.

To begin unlocking the riches hidden in these spinifex hills, Forrest just needed to win over the Yindjibarndi people. But he hadn't counted on Michael Woodley, an articulate and feisty leader of his people. Woodley, who is in his early forties, is a product of the original iron ore boom in the Pilbara. His father was a white fly-in fly-out mine worker and his mother a full-blooded indigenous woman who died as a result of alcohol abuse. Woodley attended primary school in the Pilbara town of Roebourne until Year 6, when his grandfather, a traditional custodian of the land who had sworn off the grog, took him out of the troubled town and into the bush. It was there that Woodley learned the stories and the songs of his people. "My grandfather wanted me to realise who I was, and that's a Yindjibarndi person," Woodley says. "Once I realised that, nothing else mattered anymore."

Woodley married at age sixteen and had six children, but he remained illiterate well into his twenties. His wife, Lorraine, bought him a laptop computer and he slowly taught himself how to read and write, starting with simple stories about his childhood adventures with his grandfather. "The spell-check showed me when I got the words wrong," he smiles. Woodley ended up as chief executive of the Yindjibarndi Aboriginal Corporation (YAC), the registered native title body in Roebourne. He also runs a separate organisation, partly funded by money from mining deals, which is dedicated to recording the history and culture of his people. By any standard, Michael Woodley is a success story. He and Lorraine sent their children to two of Perth's best private schools, Guildford Grammar and Perth College, and two of them have so far gone on to university.

Despite the wealth generated by the mining boom in the Pilbara over the past decade, Roebourne is an enclave of Third World poverty and despair. It seems impossible that this ramshackle town is just forty kilometres from the booming centre of Karratha and the nearby port of Dampier, from where iron ore worth tens of billions of dollars a year is shipped to Asian markets. In Karratha, upmarket high-rise apartments are being erected to house Pilbara mine workers who command salaries of at least $150,000 a year. But to enter Roebourne is akin to visiting a different country. Many of the town's 750 Aborigines live in a slum known as "the village", where alcoholism, child neglect and domestic violence are endemic. Most of the indigenous people are Yindjibarndi and many live on welfare.

Alcohol consumption in the Roebourne shire is three times the state average, according to Western Australia's Department of Indigenous Affairs. Attendance rates at the Roebourne school are 50 per cent for Year 1 students, dropping to 35 per cent by Year 8 and just 13 per cent in Year 12. In the 2008 NAPLAN test, only 20 per cent of Roebourne's Year 7 students reached the agreed standard on spelling, compared to 89 per cent nationally. Many of the town's Aborigines end up dying young or are sent to prison. A government report in 2009 found the town's notorious jail housed 177 prisoners – despite having capacity for only 116 – and 84 per cent of them were Aboriginal.

The arrival of the mining industry in the 1960s helped create the cycle of hopelessness that has almost destroyed Roebourne's indigenous people. As Karratha and Dampier were built to house thousands of mine workers, Roebourne was forgotten and most of its government offices were moved to the new regional centres. The influx of single white men brought more alcohol to the area, and Roebourne's Aboriginal women were often the only source of female companionship. Tragedy and heartache were commonplace. In 1983, a sixteen-year-old Yindjibarndi boy, John Pat, was bashed by police and later died at the Roebourne lockup from injuries that included a fractured skull, broken ribs and a torn aorta. An all-white jury in Karratha found the police officers not guilty of manslaughter.

At first glance, Michael Woodley and Andrew Forrest should see eye to eye on the best way to solve the crisis in towns like

Roebourne. Woodley appears at home in the contemporary world as he cradles an iPhone and sips on a caffe latte, yet he remains steeped in traditional law. Forrest grew up surrounded by Aboriginal people and came to firmly believe they can simultaneously hold on to ancient customs and be part of the modern economy. Both men are proud sons of the Pilbara who retain strong historical connections to the land. But the day that the mining entrepreneur came knocking on Woodley's door started a gut-wrenching tussle that has torn apart Yindjibarndi families and created a public relations nightmare for Forrest.

Fortescue initially offered the Yindjibarndi a capped cash payment of only $1.5 million a year in exchange for their support for the Solomon mining hub, which will generate a staggering $8 billion worth of iron ore a year if prices remain strong. The $1.5 million offer, spread between 1000 people, amounted to an annual payment of just $1500 per person. Fortescue later increased its offer to $4 million a year, as well as $6.5 million worth of unspecified jobs, training and housing benefits. But Woodley rejected the offer, arguing that it was insulting. Nobody disputed that the cash component of the offer was extremely low by mining industry standards. Rio Tinto, for example, pays an uncapped royalty of 0.5 per cent of revenue to other Pilbara indigenous groups; Fortescue's offer to the Yindjibarndi worked out at a fraction of that.

But Forrest is adamant that he will never throw big pots of money at impoverished Aboriginal communities as other companies have done, describing such payments as "mining welfare" that will do the same damage as government handouts have done. Providing jobs to indigenous people, he says, is the best way to break the cycle of hopelessness.

Forrest recalled the pride and dignity of the Aboriginal people he knew as a boy at Minderoo, before welfare "destroyed" them by replacing the motivation for hard work with "sit-down money".

Almost all of his mates from that era have since died. The origins of this tragedy, Forrest argued, could be found in the Aboriginal equal wages case of 1965, which ultimately led to most indigenous station hands leaving the land, including at Minderoo, because pastoralists could no longer afford to employ them. As a result of the High Court judgment, Aboriginal people across the north were forced into unemployment and off their country into virtual refugee camps on the outskirts of towns like Roebourne.

Forrest treads a fine moral line by appearing to defend the practice under which Aboriginal stockmen and servants provided mostly unpaid labour to Australia's pastoral industry. "Suddenly, you don't have an economic model," he said of the impact of the equal wages case. "All the Aboriginal people had to go, which was heartbreaking. I remember it at Minderoo ... it was very, very upsetting. They all move into town, Canberra edicts that they now have not got the salary, so they must have full benefits and equal drinking rights. Bang! The stolen generation has got nothing on this."

Forrest, as usual, has an evangelical belief that he is right about how to end indigenous disadvantage. And he is primed for a fight. He has referred to a "welfare lobby" that is "powerful" and "nasty" and argued that the biggest problem he faces is "do-gooders" who want to throw more money at the problem. He told the *Australian* in 2008 that he seethes at "idiot, feelgood white Australia, which has their beautiful but misguided equal rights program for everybody but they don't leave Toorak or Darling Point. I don't have to argue with anyone anymore; they know it simply hasn't worked."

Forrest's attitude to indigenous Australia reached a crossroads a few years before he met Michael Woodley. During a trip to the Kimberley town of Halls Creek, in Western Australia's far north, he walked the streets at 3am with his friend Herb Elliott. The pair watched in horror as local Aboriginal men rolled around drunk or stoned, fighting among themselves and assaulting women. Scores of children were milling around the streets, even at that hour. Forrest claims some of the girls offered sex in exchange for a cigarette. "When you hear 'smoke for a poke', you know you are dealing with the depths of depravity of a society which is completely broken," he said. "I found it incredibly moving that the women would scream in a way that pierces the night … I still wake up cold and sweaty thinking about it."

When Forrest returned to the scene several hours later, he found one elderly man whose leg had been run over by a car. When he tried to help the man, he was threatened with a broken bottle. Forrest then saw some other Aboriginal men he knew from the Pilbara who were headed to the bottle shop: "They bought the most powerful liquor which their money could buy, flagons of port, and I knew that the whole saga was going to repeat itself."

Soon after this, Forrest had another watershed moment while attending the funeral in Karratha of his boyhood mate Ian Black, the son of the legendary Minderoo head stockman Scotty Black. The pair, of similar age, attended primary school together at an indigenous hostel in Carnarvon, where Ian would help fight off the bigger kids who picked on the weedy Forrest because of his stutter. Forrest recalled: "Ian Black was the prized son, the son with all the get-up-and-go, the flashing bright eyes and white teeth and overt personality and charm and intellect. His sisters and his aunties explained to me what had happened to Ian over the intervening period where I'd gone to university, got a job, stayed in touch with him a little bit but eventually lost touch. He just went the welfare cycle and eventually died a premature death. The ability for me to make a choice about what I'm doing was removed by this experience. I now have no choice, I must do what I'm doing, which is to try and remove the disparity through opportunity and responsibility, because we'll never do it with welfare for able-bodied and able-minded people. In fact, welfare will remove their able body and mind eventually."

Forrest's genuine concern for the plight of Aboriginal people drove him to personally champion the hiring of indigenous workers at Fortescue. By mid 2013, more than 12 per cent of the company's employees were indigenous and the company had announced plans to increase this to 15 per cent by 2015. It's a strong record, even in the mining industry, which has been at the forefront of hiring Aboriginal people. But this is not simply corporate charity. The miners badly need workers and it's good for the bottom line to tap in to such a large pool of available labour, especially one living so close to the mines.

The mining industry's radical change in attitude towards Aborigines stemmed from the High Court's Mabo decision in 1992 and the subsequent passage of the Native Title Act. The law did not give indigenous groups the power of veto over mining projects, but they finally had a seat at the table and the right to negotiate for economic and social benefits in exchange for access to their traditional lands. Until then, the mining companies had a shameful record of discriminating against, or simply ignoring, the Aboriginal people on whose country they operated.

While many mining companies have quietly gone about boosting indigenous job numbers, Forrest has been proudly shouting his own record from the rooftops. This has particularly irked Rio Tinto, which realised its errors in the 1990s and has since become the largest private-sector employer of indigenous people in Australia. Fortescue's promotional videos feature Aboriginal employees talking about the profound impact on their lives of having a job, often for the first time. Some had been in jail, others were hooked on drugs, and most had never finished school. After graduating from Fortescue's training course and being placed in a job, they now earn salaries of more than

$100,000 a year. Every new Aboriginal recruit is handed a fishing rod as a symbol of Forrest's much-repeated aphorism: give a man a fish and you have fed him for a day, teach a man to fish and you have fed him for a lifetime. To some, this seems patronising, but to others it is clearly empowering.

Forrest was particularly touched by the story of Kris Dixon, a young Aboriginal man who joined Fortescue's training centre in Port Hedland in 2011. Dixon had grown up in a violent home in the northern suburbs of Perth, and soon fell into a life of heroin addiction and burglary. He spent almost all of his twenties in jail and lost contact with his two children. But after being released from prison and moving to the Pilbara to start a traineeship as a plant operator with Fortescue, he had firm plans to earn a good income, own his own house and reunite with his kids. "Working makes me feel like I've got structure in my life – it makes me feel good about myself," Dixon said. "I've learned from my mistakes and I'm proud of myself because I don't think I will be that person again."

Several months into his training program, however, Dixon, 32, went to Broome for Christmas and ended up in a drunken fight outside the town's nightclub. He was stabbed in the chest, collapsed to the ground and was set upon by a group of men who repeatedly kicked him. Sometime after that he died. A French national was convicted of his murder in 2013. Forrest had come to know Dixon well and was devastated by his death. In a video produced for Fortescue just before his death, Dixon had delivered a haunting message about how employment could help save indigenous people like him. "We don't want to be left behind," he said in a gentle voice. "We've probably been left behind too much. We'll be the sleeping giants, us Aboriginal people."

By August 2013, under Forrest's direction, Fortescue had awarded $1 billion in contracts for its mines to Aboriginal businesses, a move some believe is helping to create an emerging breed of indigenous entrepreneurs. The 102 contracts, awarded to fifty-two Aboriginal businesses, cover services such as catering, cleaning, waste management and camp administration across Fortescue's operations in the Pilbara. To be eligible for a contract, a business had to be at least 25 per cent owned by an Aboriginal person or group. Of the contracts awarded, more than 80 per cent went to Aboriginal businesses that were at least 50 per cent Aboriginal-owned. Brian Tucker, from the Nyiyaparli people, described winning one of the contracts as the biggest thing that had happened to his people since mining started in the Pilbara. "We want this contract to be the first of many," he said. "We want to have a future where we decide what's best for our people. We want to provide sustainable jobs to our people in our country and hopefully this will change communities, people's lifestyles and the environment."

In 2011, the company reached a novel agreement with the Njamal people of the Pilbara that will make them the managers of a mine that could one day produce 100 million tonnes of magnetite, a lower-grade form of iron ore. The joint venture will allow the Njamal to protect their cultural sites and earn revenue from providing transport, camp-building and other mining works. They won't receive royalties for the minerals that are dug up from their land, but they hope the initiative will build indigenous capacity, create jobs for their children and break the cycle of despair.

One of Australia's best known indigenous academics, Marcia Langton, was once critical of the way mining companies treated Aboriginal people. In 2010, she contended that the disparity in living conditions between towns like Karratha and Roebourne was being exacerbated by the mining boom. But by 2012, Langton had modified her thinking, arguing that the industry was leading a revolution that was empowering Aboriginal people and including them in economic life. She had also become an admirer of

Forrest and took a seat on the advisory board of his Australian Employment Covenant, the body designed to find 50,000 jobs for Aborigines.

Langton also launched a blistering tirade against "the Left" who, she said, clung to the notion of the blackfella as a "noble savage". "Whenever an Aboriginal group negotiates with a resource extraction company there is an unspoken expectation that no Aboriginal group should become engaged in any economic development," she said. "They only tolerate Aboriginal people living on their own land as caretakers of wilderness, living in poverty and remaining uneducated and isolated."

Michael Woodley has a vastly different viewpoint about the way his fellow Aborigines have been recruited into the mines. He argues that because many of his people feel such a deep spiritual connection to their country, they do not want mining jobs that involve ripping up the land. Many have other interests, including attending university and working in other industries, that they would like to pursue, but instead they are being lured into high-paid mining jobs.

Woodley is deeply critical of what he regards as Forrest's condescending belief that Aboriginal people cannot be trusted to manage significant amounts of money. He insists he will never agree to a deal with Forrest unless he pays a royalty of at least 0.5 per cent to the Yindjibarndi. With Fortescue planning to produce 60 million tonnes a year from the Solomon mining hub, such a royalty could amount to $50 million a year if iron ore prices remain high. And that is just the starting point: Fortescue has discovered more than 3 billion tonnes of high-grade ore at Solomon and plans to lift production well beyond 60 million tonnes over the coming years.

Woodley argues the Yindjibarndi are intelligent enough to manage the money for themselves. He points to safeguards that have become standard in native title agreements between miners and indigenous groups, including the use of trust structures administered by professional managers and the creation of special funds that can be drawn on only by future generations. "They think we will piss it up the wall," Woodley says. "That's a silly notion of what someone did some time back. We value how money operates and we will establish a trust structure to take care of the next generation. We have 1000 people so we need to look after them, we have to make sure they're in good health and the kids go to school. We can help ourselves."

An example of Forrest's hypocrisy, according to Woodley, is the deal Fortescue struck with billionaire Kerry Stokes's Iron Ore Holdings in 2012, which involved a cash payment of $20 million and a royalty of between 2 and 5 per cent on iron ore shipments made from one of Iron Ore Holdings' Pilbara deposits. The deal was later abandoned, but in Woodley's eyes it proved Forrest was prepared to pay industry-standard royalty rates to dig up Stokes's land while offering the Yindjibarndi a relatively minuscule amount to mine at Solomon.

Woodley also points out that billionaire Gina Rinehart receives a 2.5 per cent royalty from Rio Tinto for every tonne of iron ore mined by the London-based company in the Pilbara, a legacy of a deal negotiated by her father Lang Hancock, who pegged many of the original tenements. The agreement added hundreds of million of dollars to Rinehart's fortune and gave her the platform to start her own mining empire and ultimately become Australia's richest person.

Forrest himself has refused to allow anyone else to use his own ancestral land at Minderoo without demanding high amounts of compensation. When a gas company asked for permission to use a small area of Minderoo for infrastructure to service a nearby plant, Forrest's unrealistic demands (which ran to tens of millions of dollars, according to those familiar with the talks) sank the proposal. In another example, Forrest's private company launched a legal challenge against a sand mining company, Yarri Mining, that had applied to operate on Minderoo. Forrest argued in the WA mining warden's court that Yarri's operations would affect the viability of his pastoral lease and cause environmental damage to the land.

The ideological clash between Forrest and Woodley came to a head at a rowdy and shambolic native title meeting in Roebourne in March 2011. In the lead-up to the meeting, Fortescue had supported – and some even claim initiated – the formation of a breakaway group of sympathetic Yindjibarndi people called the Wirlu-murra Yindjibarndi Aboriginal Corporation (WYAC). The members of the WYAC had grown tired of what they saw as Woodley's domineering style and his insistence on holding out against Fortescue. They wanted to do a deal to ensure their children would at least have jobs in the mines and a roof over their heads. Fortescue sweetened the deal for the newly formed WYAC by offering the group a $500,000 "sign-on fee", as well as a $500 "sitting fee" for every meeting its members attended.

Divisions within Aboriginal communities over such issues are hardly unusual. Across Australia, Mabo has left a complex legacy of acrimony and competing interests among traditional owners that has produced winners and losers. But this dispute, which has pitted vulnerable Aborigines against one of Australia's wealthiest men, has been a particularly ugly episode. The Yindjibarndi people of Roebourne, already overwhelmed with anguish, are now hopelessly split over the best way of preserving their cultural heritage while trying to ensure economic prosperity for themselves and future generations.

Based on edited footage made public by the YAC, the Roebourne gathering in 2011 appears to be a case study in how not to run a native title meeting. With his arm in a sling after recent shoulder surgery, Forrest stood up in the town's 50 Cent Hall and turned on the charm. He explained that he was a "local boy" who had been coming to Roebourne since he was two years old, and he wanted to improve the lives of the town's Aborigines by mining on their land. "I've had one message I give and I've been giving ever since I became a businessman: the more you know Aboriginal people, the more you love them," he said.

Michael Woodley pleaded with the audience to reject Fortescue's offer, arguing they would be handing over their traditional country in return for only $4 million a year for the rest of their lives. "The Solomon hub has about 3 to 4 billion tonnes of iron ore selling at $US180 a tonne. This guy makes $6 billion a year from our country and he wants to give us $4 million cash. Now come on, we can't be that blind and stupid. This is Yindjibarndi country and we need to benefit as well. We sign this thing and we are dead."

Later, Woodley again approached the microphone at the front of the hall but WYAC's lawyer, Ron Bower, who was controlling the meeting, quickly snatched it back. The meeting descended into chaos amid arguments over how it should be run. At one point, the Yindjibarndi's most senior lawman, Ned Cheedy, who was strongly opposed to a deal with Fortescue, implored his people to be quiet and listen to him. The frail elder said later: "I remember the old people saying we have always been one Yindjibarndi people, one camp. I remember Dad saying Yindjibarndi have never split." Cheedy died a year later, aged 105, still desperately sad over the loss of his people's unity.

Attempting to wind up the increasingly rowdy meeting, Bower called for a show of hands in an attempt to gain endorsement for the Fortescue deal. Woodley's group walked out in protest and the motion was passed easily. Woodley claimed later that Fortescue had paid for hundreds of people from as far away as Carnarvon to be bused into Roebourne to back the deal. He complained the meeting was illegal because there was no traditional owner register for participants to sign to verify they were from Yindjibarndi country. But the breakaway members of the WYAC said the meeting was valid and the outcome proved they now had the right to negotiate directly with Fortescue. After the vote, Forrest returned to the front of the hall to tell the remaining Yindjibarndi he was "proud" of them for accepting him at his word. "I've seen what just money does for our Aboriginal people," he said. "If you have the opportunity to have a real crack yourself, you'll do as well as any white man every time."

Fortescue has denied doing anything wrong, either legally or ethically, in the hardline way it has dealt with the Yindjibarndi. But the company has never shied away from the fact that it was desperate to win approval to start mining at Solomon, and that it wanted to do so by spending as little as possible. In 2011, the ABC's *Four Corners* program aired video footage of an earlier meeting between the company and the Yindjibarndi at which Fortescue's land access manager, Blair McGlew, outlined his position. In what sounded like a none-too-subtle threat, McGlew explained that Fortescue would take legal action if it did not get its way because it was "in a hurry, in a rush" to develop the Solomon project. He also told the meeting that Fortescue wanted to be the lowest-cost producer in the Pilbara and it would not "pay quite the same money" to indigenous group as other companies.

After the 2011 meeting, the story of the Yindjibarndi dispute might easily have died away. But a lawyer for the WYAC who attended the meeting in support of Forrest, Kerry Savas, later resigned and claimed publicly that Fortescue had "created" the splinter group with the sole intention of enabling the miner to circumvent the obstinate Woodley and his YAC. "On the face of it, this was an Aboriginal meeting," Savas said. "However, when you look at it closely, it wasn't. The table was set by FMG and they had prepared the menu as they wanted it. They wanted to get enough people there to defeat the YAC and they did. So it is unusual when it's a corporation involved and they want to control the meeting to that extent, and to pay fees to people that are voting at a meeting where the outcome of that vote directly [affects] the company."

Fortescue, however, has always insisted the splinter group came to it for help, and that the funding it has provided to the WYAC is consistent with what occurs in the mining industry during all native title negotiations. The National Native Title Tribunal and the Federal Court have both backed Fortescue's stance that it negotiated with the YAC "in good faith", as it is required to do under the Act. By 2013, Fortescue had started providing jobs and housing to its sympathisers in Roebourne and had set up a special training facility in the town. Although it did not have an indigenous land use agreement with the YAC, the registered native title body, the company was able to begin mining iron ore at Solomon because it had secured all of its approvals with the help of the breakaway WYAC. Michael Woodley, meanwhile, was receiving nothing; Fortescue had depicted him as a man motivated only by money, who had lost the support of most of his community – suggestions the firebrand leader has strongly denied.

The messy fallout from the Roebourne meeting was highly damaging to Forrest's jealously guarded reputation as a fierce advocate for Aboriginal people. When Woodley's YAC posted the footage of the meeting on video-sharing website Vimeo, it was viewed thousands of times in just a few days. Forrest's lawyers then hand-delivered a letter to Vimeo's New York office claiming the video was defamatory and demanding it be taken down. Forrest said he received death threats after the video went viral. "The reason why it got taken off Vimeo is because it did incite racial hatred and I did receive death threats," he said. However, the video was later uploaded to YouTube and has since been viewed on that site more than 20,000 times. Fortescue responded by posting its own video on YouTube that put forward the WYAC's side of the story. The company took out advertisements directing people to the video, which it called "The true Yindjibarndi story". It has since been viewed about 8000 times.

The video of the Roebourne meeting sparked particular fury among those parts of the indigenous community that remain deeply suspicious of the motives of white Australia. The editor of the *National Indigenous Times*, Stephen Hagan, wrote in that newspaper after viewing the footage:

> It's the same old story of divide and conquer; a well tested and proven strategy of non-indigenous people when seeking to wrestle control off traditional owners. In the early contact days of British colonisation the imperialists simply identified conforming traditional owners and adorned them with shiny metal breastplates and undertook business exclusively with them when it suited. Dissenting blacks were dealt with promptly, in most cases never to be seen again.

The courts may have found Fortescue complied with the Native Title Act in its dealings with the Yindjibarndi, but the miner sailed much closer to the wind with the methods it used to win approvals for its Solomon hub under Western Australia's Aboriginal Heritage Act, a critical piece of legislation that governs whether companies can disturb indigenous sacred sites.

As the chairman and majority shareholder, Forrest was still firmly in control of the company and received regular briefings on Solomon. In 2012, it emerged that Fortescue had terminated the services of a highly experienced anthropologist, Brad Goode, after he refused the company's demands to amend sections of a report on indigenous heritage. Goode claimed Fortescue's heritage managers "tried to pressure me into altering the report" because his study backed the concerns of the Yindjibarndi about the potential destruction of several sites, including burial chambers, along the Kangeenarina Creek in the Solomon mining area. Goode had recommended a fifty-metre exclusion zone around the creek – a move that would have involved further consultations and additional approvals under the Act. He was shocked when Fortescue asked him to amend his report: "It was the worst, most reprehensible experience I've ever had as an anthropologist." In response, Fortescue said it always scrutinised the reports of its consultants to ensure their work was relevant, complete and complied with the requirements of the government. "It is not uncommon for us to raise queries and work through those with the consultant," a spokeswoman told the *Australian*.

After Fortescue severed ties with Goode, it quickly engaged another "independent" anthropological firm, MGA Consulting, to survey the same area in the Solomon hub. One of the directors of MGA Consulting is Michael Gallagher, who had worked as an anthropologist in the land access team at Fortescue for several years, resigning in late 2010 to become a "consultant" to the breakaway WYAC, the group that is funded by Fortescue. MGA's other director, and its owner, is Lillian Maher, who is also Michael Gallagher's partner. The couple share the same address in Fremantle, according to records held by the Australian Securities and Investments Commission.

Astonishingly, at the time of the survey, Maher was also serving as the state manager of the National Native Title Tribunal, which was tasked with issuing rulings on Fortescue. Further clouding the independence of the report was the fact that Maher's daughter, Lisa Maher, was employed in the senior role of heritage manager at Fortescue at the time. In her role at the National Native Title Tribunal, Lillian Maher did not declare her ownership of MGA Consulting, the work it had done for Fortescue or her family's links to the mining company. She resigned from the tribunal in August 2012.

It came as little surprise that MGA Consulting's report in September 2011 dismissed all of the concerns that had been raised by Goode about the possible destruction of Yindjibarndi heritage sites. "It is recommended that the development of the proposed works on the land proceed," said the report, which was authored by Gallagher, whose salary was indirectly being paid by Fortescue. Fortescue, as legally required, submitted both the MGA and Goode reports to the state Department of Indigenous Affairs as part of its application for approval under the Aboriginal Heritage Act.

Two months later, the Aboriginal Cultural Material Committee's specialist anthropologist, Michael Robinson, met the WYAC to discuss the glaring conflict between the two consultants' reports. The committee is housed within the Department of Indigenous Affairs and charged with advising the WA government on heritage sites. Soon after the meeting, a departmental briefing note to indigenous affairs minister Peter Collier (released under freedom of information laws) contradicted Gallagher's report, explaining that the WYAC believed "Kangeenarina Creek has a story associated and a song for it, which is part of traditions and customs central to the Yindjibarndi religious system". The note to Collier went on to state that even the WYAC believed "the creek should be protected and understood FMG had agreed to a fifty-metre buffer zone around it".

Around this time, the department was becoming increasingly frustrated in its dealings with Fortescue. A memo from senior compliance officer James Cook to his boss, director general Cliff Weeks, estimated that Fortescue and one of its earlier consultants, Alpha Archaeology, had under-reported possible heritage sites in one part of the Solomon hub by about 30 per cent. Cook said

Fortescue was unable to start work until the department had finished an investigation into its "section 18" application to disturb an Aboriginal heritage site. "This is of concern to FMG as it will hamper their operations," he wrote in an email.

Cook went on to discuss Fortescue's record in complying with the law, giving weight to the theory that the miner was more concerned about getting things done fast than following the normal processes. "FMG's compliance with the AHA has been variable," he said. "As FMG often work to tight timeframes, they often submit information related to applications under section 18 at very late notice, resulting in insufficient time being given to the department to assess that information. FMG appear to submit minimal information in relation to annual compliance reports and impacts to sites."

Despite these bureaucratic concerns about Fortescue's modus operandi, the company was ultimately able to get its way. The committee recommended that Peter Collier, as the minister, impose a condition on Fortescue's application that would prohibit any ground-disturbing activities along the Kangeenarina Creek. Fortescue appealed that recommendation to the State Administrative Tribunal, and Collier, after a private mediation, agreed to delete the condition.

Collier, a Liberal Party powerbroker in Western Australia, also happens to be a valued mate of Andrew Forrest. In 2011, the minister told a Perth radio station of his close friendship with Forrest and how he "took guidance and great advice from his wisdom". He believed Forrest had a deep compassion and interest in improving the welfare of Aboriginal people. But Collier denied that his duty to protect indigenous heritage sites would be compromised by his friendship with Forrest, who is the major shareholder of the company with the most mining tenements in the state.

When Collier deleted the condition that would have forced Fortescue to avoid burial sites and other heritage areas at Solomon, Michael Woodley was outraged. He wrote to Collier, asking whether he was "unduly influenced by your friendship and the advice you have publicly admitted taking from Andrew Forrest". Woodley said later that his people stood to lose heritage sites that were up to 35,000 years old as a result of Collier's decision. "It is a weak and morally wrong decision from Mr Collier," he said.

Anthropologist Brad Goode wasn't the only Fortescue consultant to feel as if he was under pressure to conform to the company line. Earlier, an archaeologist from New South Wales, Sue Singleton, told the Department of Indigenous Affairs that she deleted key sections of a 2010 survey of Aboriginal heritage sites in the Solomon hub because she feared Fortescue would not otherwise pay her for the work. In a letter to the department in November 2011, Singleton said she believed her firm, Eureka Heritage History Archaeology, would miss out on $70,000 that was owed unless she agreed to delete sections that referred to the Yindjibarndi's ethnographic connection to the Solomon lease.

Singleton also claimed in the letter that Fortescue had terminated her services "as they considered a 'local' archaeological firm could better provide the services (outcomes) required by them". Another former consultant to Fortescue confirms the company is known for its tendency to shop around until it finds someone who will report what it wants. "The heritage department is very focused on looking for consultants until they get one that fits with their views," the consultant says.

Fortescue has consistently denied claims that it has forced any of its independent consultants to change their reports. Development director Peter Meurs said in 2011 that Fortescue had simply requested the correction of "unqualified commentary" in the Singleton report. "We have spent many millions of dollars to protect and avoid significant Aboriginal heritage sites at the Solomon Hub and will continue to do so," he said. "From the time we begin exploration in any area, many years prior to a project commencing, we work closely with Aboriginal land owners to identify and protect Aboriginal heritage sites."

Forrest's aggressive approach to winning land deals was honed several years before he encountered the Yindjibarndi or Michael Woodley. In 1997, while chief executive of Anaconda, Forrest struck a major agreement with native title claimants in the north-eastern Goldfields as he raced against the clock to build his Murrin Murrin nickel laterite plant near Leonora. The negotiations were arduous because the region was home to scores of overlapping native title claims and the indigenous groups were divided. Forrest worked assiduously to win over local Aboriginal leaders, even building a chapel for some of them and granting one prominent elder, Sadie Canning, the franchise to run the general store at Murrin Murrin.

Under the deal with Anaconda, the traditional owners would receive a royalty on production – reportedly only 1.6 per cent – and the company would pay an additional $1 million a year into a charitable trust. "It shows we can share and peacefully coexist on the land," Forrest said. "This gets rid of the handout mentality. It will take people off the streets and out of the hotels." Forrest bristled at inquiries by journalists over the apparently large gap between his project's forecast earnings and the promised payments to indigenous people, telling reporters it was "not a line of questioning I appreciate".

The claimants later alleged that Anaconda had not paid the $1 million into the trust as promised. Two years after Forrest was forced out of Anaconda, another Goldfields elder, Ivan Forrest, said his community felt let down and wanted to sue Anaconda. "They have promised the world and they have given us nothing," he said.

One of the plaintiffs, Maria Meredith, claimed Forrest "has a lot to answer for. Our community became completely divided. We could have used native title to our advantage had we all stuck together." Anaconda's new chief executive, Peter Johnston, was left to clean up the mess.

At Fortescue, Forrest's first big native title agreement involved gaining access to 40,000 square kilometres of land in the Chichester Range to enable work to start at Cloudbreak and Christmas Creek. But he would again become mired in controversy over what some observers saw as his divide-and-conquer tactics. In August 2005, Fortescue secured the signatures of six Nyiyaparli elders who had suddenly broken away from the broader claimant group for the area. The deal would pay them a royalty rate of just 2.5 cents for every tonne of iron ore produced, as well as sign-on fees of $400,000 each. The cash amount under the deal worked out at a tiny fraction of what other miners were paying in comparable native title deals, although Fortescue promised additional benefits, including employment on its mine projects.

But the deal signed with the six Nyiyaparli elders bore no relation to an agreement that had already been negotiated by the Pilbara Native Title Service (PNTS) on behalf of the wider claim group. A former native title lawyer, Marcus Priest, reported in the *Australian Financial Review* that the new deal removed significant heritage, cultural awareness and environmental provisions contained in the PNTS agreement. It was also signed without the elders having any legal representation and without any consultation with the broader group.

Fortescue trumpeted the newer native title deal and posted photographs on its website of a beaming Forrest with the Nyiyaparli elders. It told the stock exchange that the agreement removed constraints which might have impeded access to land that was host to 80 per cent of Fortescue's iron ore resources.

One of the Nyiyaparli elders, 74-year-old David Stock, had worked at Minderoo when Forrest was growing up and had taught the boy how to ride a horse. He was still close to Forrest's father, Don. But the day after being flown to Perth to sign the deal in Fortescue's head office, Stock and the other elders pulled out of it, claiming they had not understood the details and had agreed under duress. Stock was quoted in the *Australian* as saying he felt obligated to sign the agreement because of his personal connection to the Forrest family. "I didn't know what was going on. I feel like they made me sign – they kept calling me uncle," he said. "I've done a silly thing." Another signatory, Raymond Drage, claimed that when the elders said they wanted legal advice, Fortescue chief negotiator Harry Adams had complained about delays and said Fortescue could take the matter to the National Native Title Tribunal for arbitration. "They had us in a tight corner," he said.

Forrest rejected any suggestion of misconduct and said he felt insulted by the "extraordinary" accusations against his company given his family's close relations with the Nyiyaparli people. He claimed the six elders had "pushed their way into our offices to sign the deal because of their absolute frustration of dealing with the PNTS". Eventually, the deal with the Nyiyaparli was renegotiated on better terms and the original protections were reinstated.

More recently, Fortescue ran into another conflict with indigenous groups as part of its efforts to secure mining tenements covering 4300 hectares of land near Tom Price, in the central Pilbara. Members of the Puutu Kunti Kurrama and the Pinikura (PKKP) native title group took Fortescue to the National Native Title Tribunal in 2008, successfully proving that the company had failed to negotiate in good faith as part of its obligations under the Native Title Act. (Under the Act, a miner must negotiate for six months, but at the end of that period – regardless of the outcome – it can simply go to the tribunal to get its mining lease.)

The PKKP traditional owners argued that they had only reached a broad protocol during six months of talks but there had been no discussion of any substantive issues.

One of the traditional owners, Donna Meyer, said her people were ecstatic at the tribunal victory because they were "constantly being pushed and pulled by the whims of mining companies". But Fortescue appealed the tribunal's decision and ended up winning the legal battle against the traditional owners in the High Court. The impact of Fortescue's victory worried some indigenous groups, which believed it weakened the ability of claimants to secure compensation because miners would now be able to avoid talking about any substantive issues during the six-month negotiating period.

One PKKP elder, Nyaparu Jeffries, believed Fortescue had never made a genuine effort to negotiate. "While mining companies make billions of dollars out of our traditional county, we are still living below the poverty line," he said. "FMG never began any substantial negotiations towards an agreement with our people; they just went through the motions. It makes our people sad that a company that promotes itself as helping indigenous people was putting a case to the courts to make us powerless in our own country."

The PKKP eventually ended up with a land access deal, but, like Fortescue's other native title agreements, it involved minimal monetary compensation. For the traditional owners, the process of negotiating with Fortescue had been "extremely distressing" and they believed that mining on their traditional lands would have a detrimental impact on their way of life, culture and traditions.

As these examples illustrate, Forrest appears to have subverted some of the original intention of the Mabo ruling. The High Court found in 1992 that native title is a property right – in the same way that other forms of property rights are held by millions of Australians. In other words, Aboriginal people should be entitled to receive fair compensation for access to their traditional land. Yet Forrest has come to view the native title system as a useful vehicle for him to deliver a range of social benefits to indigenous people, rather than as the recognition of a legal right to monetary compensation.

Forrest's efforts to improve the lives of Aborigines are beyond reproach, but his demand that his charitable initiatives become a condition of Fortescue's land access deals is highly contentious. His opposition to throwing large sums of money upfront at Aborigines might be understandable, but he has never explained why Fortescue cannot structure its payments so that money is put into a trust and can be used only by future generations when mining has ended on their land.

Forrest's critics also point out that wages paid to Aborigines for their labour should not be seen as a legitimate form of compensation for native title. "It amazes me to hear some people in the mining industry suggest that wages are a benefit for native title holders," said Griffith University professor Ciaran O'Faircheallaigh. "They are payment to people for their work. You can't use them to compensate a community, you can't use them to foster culture, you can't use them to set up new businesses, you can't use them for educational scholarships."

Simon Hawkins, Forrest's old sparring partner who is now chief executive of the renamed Yamatji Marlpa Aboriginal Corporation, says negotiating with Fortescue over the years has proven more difficult than with other miners because of the company's insistence on forcing its own agenda on indigenous groups. "They want to force upon indigenous people their own values associated with what they consider welfare," he said. "Other companies tend to have a more sophisticated approach that balances commercial interests with legal obligations and corporate social responsibility, which ultimately creates stronger agreements for traditional owners."

Forrest's policy of making lowball offers and driving hard bargains also risks overshadowing much of his excellent philanthropic work. And it has given his critics ammunition to argue that his mission to get Aborigines into work is driven by his own commercial motives. Some also say his championing of the indigenous cause is just part of another slick Forrest campaign to boost his public image and cosy up to governments.

Michael Woodley, one of his fiercest critics, says people should see through Forrest's crusading style and realise that the only reason he wants to be seen as supporting the Aboriginal cause is because he wants to mine their land. "If he didn't have a mining company, we wouldn't even fall on his radar," Woodley says.

Plenty of other Aboriginal leaders, however, are convinced that Forrest's approach, while not perfect, is generally the right one. Clinton Wolf is a former AFL footballer who has worked on both sides of the fence in Western Australia: first as the head of a major native title representative body in the Pilbara and now with a Perth mining contractor that has won contracts from Fortescue. He says people initially questioned Forrest's plans to boost indigenous employment and deliver contracts to Aboriginal firms, but he has delivered on all his promises while managing to bring an important issue to national attention.

Wolf, 43, also knows from first-hand experience that the influx of mining money in the Pilbara hasn't solved any of his people's deep-rooted problems. "When I was a kid, everybody worked and all the kids, black and white, were at school," he recalls. "Everyone had pride in themselves and carried themselves in a dignified manner. I look at the amount of money that's been paid out in royalties throughout the Pilbara and I'm struggling to see any improvement in lifestyle. Where you do see improvement is where people have got jobs and are earning good money and are feeling like they're worthy. Their kids pick up on that and say, 'That's what I want to do when I get older.' Now, that's improvement."

Only his harshest detractors accuse Andrew Forrest of not being genuine in his desire to lift the lives of Australia's indigenous people. Those who know him are convinced he is sincere. But it's clear the methods Forrest has employed at Anaconda and Fortescue in trying to help indigenous people have left an unnecessary legacy of bitterness and division.

As with everything else he has done in life, Forrest's strident views and sense of righteousness have often inflamed rather than healed. Yet he sees bravado and force as the only proven methods of achieving the impossible. It's how he built two mining companies from scratch, and it's how he believes he will solve Australia's greatest moral disgrace.

## 14.

# DOING GOD'S WORK

*I have no wish to die rich.*
—Andrew Forrest

When Andrew Forrest signed a public pledge in 2013 to give away most of his fortune to charity – a promise that could make him Australia's greatest philanthropist – he revealed the biggest inspiration for his giving: the Bible. Forrest's devout Christianity is unusual for a prominent Australian entrepreneur, perhaps especially so for one who thrives on risk and adopts a win-at-all costs approach to business. But none of Forrest's friends doubt his piety. They have experienced enough Forrest-led prayer sessions, or caught him sneaking a quick read of his Bible during enough hectic overseas business trips, to understand that belief in God appears to guide his every waking hour.

Forrest "found" God as a nine-year-old boy in the sandhills near Minderoo. He grew up in a family of Anglicans, but the Forrests were not overtly religious. Late one afternoon, Andrew was riding his motorbike miles from the homestead when he decided to throw the key away in the dunes to see if he could find it again. As he now relates the story, he couldn't find the key despite several hours of frantic searching and was preparing to brave the elements as sunset approached. As a last resort, he decided to pray. Miraculously, the key turned up right in front of his eyes.

To non-believers the story might sound faintly absurd, but for the young Andrew Forrest it was a seminal event. He relayed the anecdote in 2012: "One of the reasons I became a Christian and am still a Christian now is because after about three or four hours of fruitless search amongst the spinifex and the hot sand for this key which I was probably never going to find, I tried everything, every physical thing to find the key. And eventually when I had given up all hope, I knew that Mum and Dad were going to come out in their four-wheel drive and probably start to search about sundown and get to me about midnight, and my hide would be quite properly and fairly tanned for throwing away the key to a motorbike. I prayed. And when I came up from praying, there was the key between a little petrol overflow hose and the carburettor on the motorbike. Now there's not a snowflake's chance in hell that I threw it there; I threw it over my shoulder. So I just thought okay, that's cool, we're showing off a little here, God. That's a very obvious sign, I won't ever test you again – and I haven't."

As an adult, Forrest has never hidden his religious beliefs. When he opened the Murrin Murrin nickel plant in 1999 alongside prime minister John Howard, he asked the 300 guests to bow their heads as he recited a lengthy prayer. And in 2012, after beating civil charges laid by the Australian Securities and Investments Commission, he told a gathering of friends and colleagues that he had triumphed because God was on his side. Chief executives of major companies will usually reel off a long list of well-known business texts when asked for their favourite books. But Forrest says all the guidance he has ever needed as an entrepreneur is derived from the Bible. He told a Christians in the Marketplace function in 2010, "I get all the inspiration I ever need just from reading the New Testament. That has so much [on] how to run businesses and how to work with people … I don't have to go to all those other texts. It's really all there for me."

Despite the encounter with God as a kid near Minderoo, none of Forrest's old friends recall him being overtly religious as a teenager or a young adult. His piety seems to have flourished only after he met Nicola, who hailed from a devout Christian family and had a moral compass that Forrest needed badly at the time. "He wasn't really religious until he got married," says one old friend. "Nicola is a big part of why Andrew has embraced Christianity."

The Forrests' faith in God was fortified by the stillbirth of their baby girl, Matilda, in the late 1990s. Nicola's pregnancy had progressed smoothly and there were no outward signs that anything was wrong. Matilda would have been the couple's third child and a baby sister to Grace and Sophia. The impact of Matilda's death on Forrest was profound, although few of his colleagues at Anaconda Nickel knew of his private turmoil. Forrest took counselling from pastor Ken Drayton at the St Philips Anglican parish in Cottesloe and began attending church more frequently.

These days, Forrest's local priest at St Philips is Malcolm Potts, whose Sunday sermons are unconventional and laced with humour, delving into issues such as sex and wealth. In one of his recent homilies, Potts discussed his belief that once someone is earning more than $70,000 a year, having any more money won't make them any happier. Potts's personal definition of rich is having the ability to go into the local supermarket and buy whatever he needs without worrying that he'll have enough money at the checkout. "All but a few of us are rich by world standards," he told his well-heeled congregation.

Potts's richest parishioner, Andrew Forrest, wasn't in church that particular Sunday morning for the sermon on wealth and happiness. But if he had been, it's unlikely he'd have been squirming in his pew. Because these days, Forrest's opinions on money – or at least those he has espoused publicly – are very similar to his priest's. Forrest says his greatest pleasure as a billionaire is derived from giving money away to those less fortunate than him. Friends say these views were actually formed long ago through the prism of his religious convictions. Nicola even told one interviewer that the couple liked to give their money to Christian-based activities "because that is what gives life to people".

Forrest's views on wealth were entrenched by 2007, when his fortune on paper had already hit $4 billion. He told Perth journalist Mark Drummond that his three children knew they would never have to worry about the "burden" of a huge inheritance. "What all my kids know is that they're not going to inherit it and they're happy about that. I have no wish to die rich, and my children have even less wish to inherit great wealth. We are not about creating a Forrest dynasty, we're about helping others. If you're pursuing wealth for the pursuit of wealth's sake, you're not going to die happy. And the prospects of keeping your family together, and yourself whole and happy, will be very limited. I can understand public interest in wealth, but I also admire those people like Warren Buffett and others who use their wealth to help others. If I've ever found intense satisfaction out of wealth, it's the ability to help others with it."

A few weeks after these comments, Forrest embarked on the first stage of giving away chunks of his fortune. He called a media conference in Perth to announce that he would give 1 million of his shares in Fortescue and 115 million options in a nickel mining company he chaired, Poseidon Nickel, to the Australian Children's Trust (ACT), the charity for indigenous and other underprivileged children that he and Nicola had set up several years earlier.

The combined value of the Fortescue shares and Poseidon options at the time was about $85 million, making it the single biggest act of corporate philanthropy in Australian history. Forrest said he wanted to follow in the footsteps of the modern American philanthropists like Warren Buffett and Bill Gates, and he hoped his generosity would inspire others in Australia. "If this acts as an accelerant in the thinking of others, of course I'd be deeply honoured," he said.

Forrest also gave another 520,000 of his Fortescue shares – worth more than $20 million at the time but soon to be worth more than twice that – to his brother, David, and sister, Janie. In total, he had given away more $100 million in one day.

The $85 million in shares Forrest donated to the ACT was a generous gift. But it didn't end up being anywhere near as generous as Forrest had intended. As leading financial journalist Neil Chenoweth would later document, Forrest had saved himself a huge tax bill by electing to have the 115 million Poseidon options issued to his charity rather than to himself. "If the options had been issued to a Forrest company, it would have triggered a tax bill of at least $12.9 million at a company tax rate," Chenoweth explained in the *Australian Financial Review*, adding that senior accountants had backed his methodology.

Similarly, Forrest's donation of his Fortescue shares to the ACT was, in the end, not quite what it originally seemed. The gift of 1 million shares was made through the Forrests' family trust, the Peepingee Trust, and had a value at the time of $42.6 million. Within a few months, the value of the donation had soared and the charity was flush with cash. But a year later, the global financial crisis had arrived and the value of the shares held by the ACT had slumped to $26.8 million (the Poseidon options were worth nothing to the ACT because the share prices had fallen below the 40 cent price threshold for exercising them). The donation by the Peepingee Trust would have produced a $42.6 million tax deduction for Forrest, according to Chenoweth's calculations. At a 30 per cent tax rate, the tax benefit would be $12.8 million.

Forrest's combined tax savings from the share gift and options issue came to $25.7 million. Forrest had made an $85 million donation but, thanks to tax benefits and the falling sharemarket, the net cost of the donation to him was an estimated $1.1 million. "The ACT did indeed receive a huge donation, but the bulk of the wealth given to the ACT to spend on indigenous welfare was paid for by taxpayers," Chenoweth reported. Of course, Forrest could not have known in 2007 that the global financial crisis would radically shrink the value of his donations.

One of Forrest's previous charitable donations, however, was closely scrutinised by the Australian Tax Office. When Anglo American and Glencore removed him as chief executive of Anaconda Nickel in 2001, Forrest came to an agreement with the board for it to donate $3.5 million to a company called Leaping Joey, the trustee of the ACT, in lieu of a redundancy payment. As Forrest would later testify in a court statement, he arranged the payment to Leaping Joey because he wanted something positive to emerge from his bruising fight with the two foreign companies. "As I was driven into work I kept thinking about how I could make some good out of this disaster and this treachery," he said.

But what appeared to be a noble gesture lost some of its shine when it emerged that the ACT had used the money to buy 7 million Anaconda shares from Forrest. Moreover, the charity bought the shares for 85 cents each when the closing price on market was 74 cents. Forrest pocketed $3.5 million from the deal, although he also gave the charity an additional 2.9 million Anaconda shares for free. Five weeks after the deal, in January 2002, the debt-laden Anaconda announced a $457-million loss, and by mid 2002 its share price had sunk to 29 cents. Forrest had given his charity $5.6 million worth of shares, which were suddenly worth only $2 million.

At that point, nobody was complaining. Forrest could claim a hefty tax deduction on the share sale and the ACT had $2 million of fresh funds in its coffers – although most of it was courtesy of taxpayers. But the Australian Taxation Office took a very dim view of the scheme, ruling that Forrest was trying to avoid paying tax. The taxman viewed the $3.5 million from Anaconda as a termination payment to Forrest rather than a deductible charity donation. Forrest challenged the ATO ruling, but both the Administrative Appeals Tribunal and the full bench of the Federal Court agreed that the payment was subject to tax.

Just why a charitable foundation dedicated to helping underprivileged kids was buying shares in a struggling nickel miner has never been fully explained. When the media revealed the nature of the deal, Leaping Joey said in a statement that Forrest had no input into the charity's decision to buy his shares. But that claim was puzzling to many, given that ASIC documents showed that on the day Leaping Joey bought the shares in January 2002, its only board members were Andrew and Nicola Forrest. An amendment lodged with ASIC about a month after the purchase listed Perth communications consultant Valerie Davies and Goldfields indigenous leader Sadie Canning as having been directors at the relevant time. The amendment stated that Nicola Forrest had resigned from the board at the time of the purchase. These were "administrative oversights", according to company secretary Walter Staniforth. When news of the transaction became public, Davies resigned from the Leaping Joey board within days. But Canning was later quoted defending Forrest: "Don't we all do something for our personal tax gains?"

The Anaconda share deal was vintage Forrest: it may have been devised with good intentions, but it polarised opinion over his real motives and gave his enemies another opportunity to attack him. When news broke of the deal, Forrest said he was hurt by criticism that he was trying to dodge tax and claimed he would have been better off financially if he had simply taken the $3.5 million as a redundancy payout.

His detractors didn't believe him. Warwick Grigor, his disgruntled former business partner, went on national television to accuse Forrest of hypocrisy. "The donation that Andrew sought to have put into Leaping Joey is how he'd like to be seen, as a philanthropist doing good for kids," Grigor said. "But the fact that it bought shares off him is just more evidence that Andrew's always got a hidden agenda and he doesn't do anything totally ... out of the goodness of his heart."

Forrest despises any media coverage that raises these sorts of questions about his philanthropy, claiming it will discourage others from donating. "If we were too sensitive, these recriminations would have inhibited our giving," he told *Australian Philanthropy* magazine in 2012. "The media have a serious role to play here. If the press fosters this negative attitude, philanthropy will be kept behind a protective wall of privacy, and people won't want to give openly, and as a consequence are likely to give less than they possibly could. However, if philanthropy was celebrated, then we'd have an Australian psyche which would encourage giving of all types, openly and privately."

Forrest didn't let the critics stop his efforts to help address Aboriginal disadvantage. In fact, by the time he was running Fortescue in 2007, he had decided to spearhead a movement called GenerationOne, which he boldly predicted would eliminate indigenous disadvantage in Australia in just twenty years. Forrest began talking to his fellow captains of industry about setting up a joint corporate and government pact that would create thousands of guaranteed jobs for indigenous people.

At the core of Forrest's philanthropic work was his religious faith; Andrew and Nicola even chose to call the indigenous employment pact a "covenant" because of the word's strong biblical connotations. Forrest had spearheaded the hiring of indigenous workers on his own mines at Anaconda and Fortescue, but that had also made sound commercial sense. In contrast, his crusade to end the disparity between black and white Australia was inspired by the deeds of Jesus Christ.

Forrest contacted three leading indigenous figures who had long opposed passive welfare in Aboriginal communities – Noel Pearson, Marcia Langton and Warren Mundine – and went to work on convincing the newly elected Rudd government to back him.

In August 2008, Forrest huddled with Kevin Rudd and his business adviser, Sir Rod Eddington, on the sidelines of a conference in the Queensland resort town of Coolum, trying to sketch out the details of the covenant. For Rudd, the attraction of the plan was obvious: Australia's richest person (at the time) would persuade his corporate mates to create 50,000 jobs and the government could share in the glory of having reduced indigenous unemployment and poverty.

Observers say Forrest and Rudd began talking with the modest, perhaps sensible, aim of initially creating 5000 jobs for Aborigines, but they quickly started bidding each other up. They decided 5000 jobs just wouldn't "cut through", so Forrest suggested starting at 10,000 or perhaps even 15,000. By the end of the chat, they had decided 50,000 jobs in two years would grab people's imagination. Rudd and Forrest announced the pact at a press conference the next day, alongside Pearson, Mundine and Eddington. Pearson was thrilled with the initiative but startled by the ambitious target. "It was a complete hit to the solar plexus when Andrew proposed not a few thousand real jobs in a timeframe, but 50,000 guaranteed real jobs," he said.

Forrest's initial promise sounded unambiguous: companies that signed on to the covenant would create a total of 50,000 permanent full-time jobs for Aborigines, and the jobs would all exist within two years. Rudd pledged that the government would fund and provide the training so participants in the scheme were ready for employment. At the end of the press conference, Forrest gathered Rudd, Pearson, Eddington and Mundine into a semi-circle for a photo opportunity and told them to place their hands on top of his. When they did, he said: "This is the Australian Employment Covenant" (AEC). It was front-page news the next day.

Like all of his hugely ambitious ventures, Forrest had to contend with naysayers. Professor Jon Altman, a leading academic in indigenous policy at the Australian National University, suggested a target of 50,000 jobs was "verging on being slightly dangerous rhetoric" and wondered whether the details had been properly thought through. Altman pointed out that Australia currently had a total of 50,000 indigenous people in full-time private-sector work, yet Forrest and Rudd were promising to double that number in only two years. He also warned that only 25,800 indigenous Australians were classified as unemployed across the nation. A further 129,700 were outside the labour force, many of whom lived on traditional lands, had poor levels of education and would be difficult to lure into paid work.

Forrest won support for the covenant from his billionaire mates James Packer, Kerry Stokes, Lindsay Fox and Frank Lowy. But it took him only a few months to start backtracking on his promise. First, he quietly dropped the two-year deadline. Later, the central pledge to create 50,000 new jobs transmogrified into a vision to secure 50,000 job "pledges" from employers.

In 2010, when the original two-year deadline arrived, only 2800 jobs had been filled. However, 20,000 jobs had been "pledged" by employers. Forrest began blaming the Rudd government for failing to uphold its end of the bargain to train people to fill the jobs, claiming Rudd had not followed through on a promise to "move the bureaucracy" to ensure government agencies trained workers for specific roles.

Forrest had adopted his tried-and-true corporate tactic of achieving the impossible by setting unrealistic targets, but this time it hadn't worked. Still, he had managed to bring a critical issue to the attention of corporate Australia. In a study into the scheme in 2010, researchers at the Australian National University found that although Forrest's claims about the AEC's progress had been "overstated", the 20,000 job pledges were still a "significant" achievement and represented a notable shift in the landscape of indigenous employment.

When he appeared on the ABC's *Q&A* program in 2010, Forrest flatly denied ever promising that the jobs would be filled in two years and he blamed the government for publicising the figure. But journalists at the Coolum press conference in 2008 all reported him as discussing the two-year target. Forrest also went on radio a day after the announcement of the AEC to say he believed it could be done in two years, although he did warn it was a difficult target. "I've shared with the prime minister privately that we would have crack at this as an internal target within two years," he said. "But I would say to you, that is a very steep target, and if we did it in longer than that, I'd think the result would be outstanding. We would have changed, as an Australian nation of employers, the course of social history for our indigenous brothers and sisters."

Despite the missed targets, Forrest has not given up. By 2013, the AEC had more than 60,000 job pledges from 300 employers and had placed more than 14,000 indigenous people in work since the start of the scheme five years earlier. The AEC also reported a 70 per cent retention rate for participants after six months, compared with the government's Job Services Australia retention rate of 45 per cent after three months.

These are impressive numbers by any measure, and Forrest deserves credit for overseeing the AEC's push to get thousands of indigenous people into work. There is doubtless merit to his argument that red tape has hindered his original, outsized vision. Yet by the middle of 2013, the AEC appeared to be stalling. A scan of the organisation's website at that time revealed only forty-four available positions – and thirty-nine of them were roles at James Packer's Crown Casino.

Warren Mundine, the Aboriginal leader who was involved with the AEC in his role as head of Forrest's GenerationOne until his resignation in May 2013, believes the biggest flaw in the scheme is the government's inability to train people to meet the specific needs of employers. "The problem is structural within the Department of Indigenous Affairs – it's the way it's been operating for the last thirty to forty years," he says.

Mundine is full of praise for Forrest's activism and the business nous he brings to solving problems. When Mundine took a break in 2012 to have heart surgery, it was Forrest's enthusiasm for the indigenous cause that motivated him to get back to work as quickly as possible. "I just wanted to drag myself out of bed and do the job, even when I was in hospital," Mundine says. "When you have a conversation with him, you get caught up in all that passion and you just want to go out there and rip arms off just to make it successful." Mundine didn't enjoy the phone calls from his boss at all hours of the day and night, and the pair clashed on some issues. But he came to be convinced that Forrest genuinely wants to help Aboriginal people. "A lot of people say all sorts of things about Forrest," he says. "But he could have been just a Forrest – the Forrests in Western Australia are pretty famous people. He could have been just a billionaire. But he's taken this passion and put it in this area. My belief is he has a very genuine and strong heart, but with a business brain."

Both Forrest and Fortescue are highly sensitive to criticism over their indigenous jobs record. An Aboriginal public servant in Perth, David Collard, who coordinates indigenous engagement in the state's Natural Resource Management program, was quoted in a newspaper report in 2012 as saying that Forrest's push for indigenous people to work in the mines was misguided because mining offended the traditional culture of Aboriginal people. He said Aborigines would prefer to be involved in rehabilitating the land rather than digging it up. Soon after the report appeared, Collard received a phone call from a Fortescue public relations executive who wanted to meet to discuss his views. When he turned up to the meeting at a local café, the executive had brought along three other Fortescue executives to explain to Collard how the company was benefiting Aboriginal communities. Collard was heavily outnumbered, but says he held his ground. "I was blown away," he recalls.

After 2008, Forrest began to broaden his philanthropic efforts beyond indigenous disadvantage to include disaster relief. Growing up at Minderoo, where drought, bushfires and cyclones were a way of life, he felt a personal connection to victims of natural disasters. When Queensland was devastated by floods and cyclones in 2011, the Australian Children's Trust helped in the rebuilding effort by securing corporate donations, working with local businesses to re-establish themselves and providing financial support for families hit by the disaster.

This was a template Forrest had used in the wake of the 2009 Black Saturday bushfires in Victoria, which killed 173 people and destroyed thousands of houses. Forrest swung into action within days of the disaster, flying straight to Victoria and later dispatching Mal James, the head of the Australian Children's Trust, to help break through red tape and get the relief operation moving faster on the ground. The charity spent $100,000 of Forrest's money erecting a temporary community centre in the main street of Marysville, a town virtually wiped off the map.

In the weeks that followed, Forrest visited Marysville several times and even managed to bring some razzamatazz to the huge relief operation. He organised a charity golf day with Greg Norman that raised close to $1 million and ended with a surprise concert by American music legends Simon and Garfunkel. On another occasion, Forrest walked the blackened streets of Marysville with Hollywood actor Cate Blanchett, whom he described as a "mate".

Forrest said he was encouraged by the amount of money and goodwill pouring into the affected area at a time when the Australian economy was in the doldrums as a result of the global financial crisis. "In a recession, in a smashed economy, you've got individuals that turn a blind eye to where the global market is going and make a commitment to the people," he said. "That is the most beautiful gift of being an Australian."

Marysville residents later reported how impressed they were that Forrest devoted such a large amount of his time to the cause and played a major role in reconstruction efforts. It was an extraordinary effort from a man who was also running a major listed mining company, which at the time was suffering badly amid the global downturn. "Forrest is one of the people who have hung in there," Marysville community leader Doug Walter said a few years later. "I could give you a catalogue of people who turned up to get their photo in the paper, made extravagant promises and then disappeared, but he's certainly not one of them. He came in with this really positive message: that there will be a tomorrow, there is hope and we will be able to rebuild our lives. He captured the hearts and minds of everybody involved. He's such an enthusiastic man and it's catching."

Walter recalled that Forrest arrived on a helicopter in his paddock. "He told us to take a ride in the chopper and have a look at the burn. We left him and his daughter with this elderly couple in their eighties who had lost everything. And when we got back we found him kneeling on the floor in front of them, holding their hands, crying." A local plumber, Bruce Ackerman, had a similar tale of Forrest's empathy: "I had one of Australia's richest men help carry the groceries into our kitchen a week ago and then sit down and have a coffee and actually listen to what it is we need to get the place back up on its feet."

Forrest's most symbolic act was donating 1000 "dongas", or small transportable huts, which had been used in his Pilbara mines, to serve as temporary accommodation for homeless Marysville residents. The Victorian premier at the time, John Brumby, recalls that Forrest had asked the government what was needed and was told that emergency accommodation was a priority. The Fortescue dongas were immediately trucked across the Nullarbor by Lindsay Fox's transport company, Linfox. Brumby says Forrest was a magnificent supporter of the relief effort and one of the few non-Victorians to become closely involved with it. "I couldn't fault him," Brumby says. "He wasn't a Victorian and he wasn't looking for anything from our government."

Some of the people working below Brumby, however, became irritated at Forrest's apparent skill in stage-managing appearances that ended up on the nightly television news bulletin. And there were people in Marysville who, while not wanting to appear ungrateful, complained that Forrest's dongas were designed for single men working on mine sites and were too small and dark to inhabit for any length of time. A prominent Marysville community leader says while Forrest's generosity was appreciated, many of the 1000 dongas were never used because they were the wrong sort of accommodation for families and did not contain kitchens, although the Anglican Church later provided each one with a microwave oven.

But the real shock to those working on the rebuilding mission came when Forrest sent a bill for $300,000 to the Victorian government to cover the cost of returning the dongas to the Pilbara. The government had agreed to pay for the transport costs as part of the original contract, which was signed hastily after the disaster, but many working on the reconstruction were unaware of this and had assumed Forrest would pick up the tab. "The Victorian taxpayer was left with a bill for about $300,000 to send them back," says the Marysville community leader, who has asked not to be named. "They ended up costing taxpayers, and they weren't even effective."

The Marysville leader, who worked closely with Forrest, says the billionaire's efforts were impressive, but they were relatively minor in a reconstruction operation that ended up costing more than $1 billion. "Twiggy's a very complicated individual," he says. "He's not just all show – he is motivated by good values. But he doesn't have the self-awareness to understand the impact of his actions."

Since the Victorian and Queensland disasters, Forrest's philanthropy has gone into overdrive. Nicola became chief executive of the Australian Children's Trust in 2010 and, in effect, the driving force behind Forrest's image as a humanitarian. Barely a month goes by without a new announcement. The couple have donated millions of dollars to the Salvation Army and St Vincent de Paul and Forrest has slept rough several times as part of the latter group's annual CEO Sleepout. The family's money has also been poured into dozens of charities, including Mission Australia, a Christian organisation that helps needy Australians; YouthCare, a multi-denominational group that provides chaplains in schools; and Fairbridge, a charity working with young offenders.

In 2011, Forrest gave $3 million in Fortescue shares to the Art Gallery of Western Australia and another $3.7 million worth of stock to be divided among the state's four big performing arts companies. Later that year, he gave $1.3 million in Fortescue shares to Murdoch University's Institute for Immunology and Infectious Diseases – the biggest personal contribution in the university's history.

Suddenly, Fortescue shareholders included arts aficionados and academics who might not normally study the ups and downs of the resources sector. The organisations, of course, were free to sell the shares immediately and take the cash. But in Forrest's eyes, they would be smarter to hold on to the shares, take the annual dividends and watch the value of the investment grow. "They [the shares] are dividend-creating, they are proudly Western Australian," he said. "The [recipients] can sell them whenever they wish, but if they want to see them grow in value then they can grow along with the Western Australian industry."

Forrest's next philanthropic mission was much bigger in scope and ambition than anything he had attempted previously. So big, in fact, that some of his close friends rolled their eyes at its sheer grandiosity. In 2012, Forrest announced that he wanted to end the curse of global slavery and was forming an organisation, Walk Free, to tackle the problem. Finally, this was the issue that would put him on the world map for philanthropy, potentially touching the lives of millions of people and bringing him closer to Bill Gates, whose own evolution from business magnate to humanitarian had earned him popular acclaim. Forrest has met Gates several times in recent years to discuss global issues and appears to have modelled himself on the Microsoft founder.

Warren Mundine, who was running Forrest's GenerationOne at the time, recalls his shock when his friend told him about the anti-slavery venture. "I laughed and said, 'Mate, you're just not satisfied with fixing the Aboriginal issue – now you want to save the world,'" he says. "But I love that about him."

Forrest's interest in abolishing global slavery was sparked by a trip his eldest daughter, Grace, had made to an orphanage in Nepal a few years earlier when she was still at school. When the Forrests returned to the orphanage a year later, they found that none of the children were still there and suspected they had been sold into forced labour. Forrest managed to have the orphanage shut down, but he knew he also needed to do something about the broader problem. His own investigations revealed that an estimated 21 million people around the world were living in slavery, a practice either tacitly supported or actively encouraged by governments and major companies.

Grace Forrest went on to study international politics at university in Perth and helped convince her father to set up Walk Free. In 2012, Forrest took a big step by hiring Nick Grono to run the organisation in Perth. Grono had an impeccable pedigree, having worked as chief of staff to Howard government attorney-general Daryl Williams before moving to Brussels to take a senior role at the International Crisis Group, which was led by former Labor foreign affairs minister Gareth Evans. Forrest knew he needed someone of Grono's stature to run the new group, and at one stage he even flew to Greece, where Grono was holidaying with his family, to persuade him to return to his former home town of Perth.

Grono accepts that it is highly unusual for an organisation dedicated to a global cause to operate from a relative outpost like Perth. But he says this is not such a big issue because Walk Free is using Facebook and other social media to build an awareness campaign around the issue. By the middle of 2013, Walk Free had signed up 2 million "supporters" through its website and was on track to have 4 million by the end of the year, prompting Grono to speculate that it may be the "fastest-growing global online movement ever". Grono says Walk Free wants to sign up thousands of companies to pledge to end modern slavery by eradicating any forced labour in their operations and supply chains. Walk Free is planning to launch a global slavery index, which will be the first systematic effort to measure the prevalence of slavery and analyse countries' responses to the issue. It also plans to create a global fund backed by major donors.

But exposing the evils of slavery won't make Forrest too many friends among those companies around the world that have grown rich from the practice. Already, Walk Free has come into conflict with companies such as Spanish fashion chain Zara, US retailer Target and Japanese electronics manufacturer Nintendo. That doesn't appear to faze Forrest, who thinks big business is more likely to act when targeted personally by him rather than by a run-of-the-mill non-government organisation. "We're able to reach into the most influential families of the world and discuss with them candidly that the world has a problem," Forrest said.

For Forrest, a breakthrough of sorts came in March 2013, when then prime minister Julia Gillard – who had come to despise the businessman in the wake of the mining tax furore – publicly praised Andrew and Nicola's "amazing leadership" on fighting slavery and promised that the Australian government would take a bigger role in the campaign. Forrest had put $8 million of his own money into establishing Walk Free in 2012 and was on track to tip in a further $12 million by the end of 2013. This generosity presumably involved Forrest dipping into the $210 million in dividends he has received from Fortescue since 2011. Grono says his boss is highly dedicated to the cause, spending time in the group's Perth office but also meeting government and business leaders whenever he is overseas.

Forrest nominates his most inspiring historical figure as William Wilberforce, the politician and philanthropist who played a central role in abolishing the slave trade in the British Empire in the nineteenth century. In what may be a template for Forrest's own ambitions, Wilberforce was also an evangelical Christian who believed that a career in politics was the best way he could serve God. Forrest's favourite book is *A Man for All Seasons* by Robert Bolt, the 1960 play about Sir Thomas More, the deeply religious statesman whose adherence to principle led to his death at the hand of Henry VIII.

The rise of Andrew Forrest as the nation's greatest philanthropist culminated in 2013, when he and Nicola became the first

Australians to sign the Giving Pledge, the movement established by Gates and Buffett in the United States. By signing the pledge, the Forrests promised to give away "the vast majority" of their fortune, either during their lifetime or when they die. In a letter explaining their actions, Forrest called on other wealthy Australians to give more to charity. "With laughter we read on a physio's wall a caption, 'Those with the most toys when they die, still die.' How true," the letter said. "Guided by the same principles of the book which inspired the successful leadership of our companies, the New Testament, we chose to help those least fortunate. In our country, this is our first Australians, and globally, those suffering the unbearable yoke of modern slavery and forced labour."

Forrest has so far given away almost $300 million of his fortune. But his signing of the Giving Pledge also raises the question: how can he possibly give away most of his wealth when almost all of it is still tied up in Fortescue shares? Forrest has given away small amounts of shares as donations in recent years, but he has shown no desire to significantly reduce his controlling 33 per cent stake, which affords him the ultimate say in Fortescue's corporate strategy. It's also unclear just how much money Forrest will ultimately be able to give away. The value of his Fortescue stake has fluctuated between $2 billion and $13 billion since 2008, meaning the commitment he and Nicola have made under the Giving Pledge will always be subject to extreme volatility. While Forrest could easily be worth several billion dollars by 2020, his net worth could just as easily have plummeted.

What is driving Andrew Forrest's quest to become a philanthropist on a global scale? There are several theories. His religious faith is evidently central to his belief in trying to help those less fortunate than himself. But in recent years, his generosity has been spurred by the attacks on his so-called greed that were led by Wayne Swan in the wake of the mining tax brawl. Swan was one of the few people unmoved by Forrest's philanthropy. "Charity is not a substitute for paying tax," the then treasurer said. Swan's repeated public denigrations may have prompted Forrest to boost his charity work in a bid to win the PR battle against a Labor government he believed was intent on starting a class war.

Another theory is that Forrest desperately wants to rub shoulders with global business and government leaders like Gates, and philanthropy is a useful means of winning respect among the world's powerful elite. Such is his standing in the eyes of the establishment these days that Forrest was invited to speak on his philanthropic work before a VIP audience, including Queen Elizabeth II, at the 2012 Commonwealth Day ceremony at London's Westminster Abbey.

A former Fortescue senior executive says that while Forrest genuinely believes in the various causes he champions, he also realises that they aid his ability to cosy up to politicians and advance his own commercial interests. "He's not unaware of how philanthropic issues open doors, particularly with politicians," says the former executive. "Andrew learned very early on that if you could set up a meeting with a politician, he'd always take your meeting with you on your desire to help indigenous people. And after that little chat was over, Andrew could say, 'Now I've got these other points I'd like to talk to you about as well.'"

Perhaps there is also some "billionaire guilt" at play. Under this theory, the mega-wealthy in society experience acute pangs of guilt whenever they step out of their five-star hotels or private jets and glimpse people who are really struggling. For Forrest, this feeling of guilt, and his desire to do good, may be amplified by a win-at-all-costs approach to making money.

Warwick Grigor subscribes to this latter theory when analysing his old business partner's bid to reinvent himself as a philanthropist. Although he once believed Forrest had ulterior motives, especially after the Leaping Joey controversy, Grigor has come to admire his growing social conscience, seeing it as an important progression in his character. "I genuinely believe he wants to do good. In essence, he's got a good heart," Grigor says. But Grigor also sees a conscience that needs assuaging: "He does have to contend with a sense of guilt; you can't be a good Christian, do all that he's done and not want to recompense for that. It's only logical."

Another former business associate told the *Sydney Morning Herald* that Forrest was a man haunted by his past, particularly the mess he left behind at Anaconda Nickel. "He does not want people to think he is a bad man. That's what drives him. I don't think he ever intended to let people down with Anaconda. It went the wrong way for him as CEO and he has wanted to make it up to the universe ever since."

Investment banker John Poynton, who is at the forefront of encouraging Perth's wealthy to become involved in philanthropy, prefers to take a sanguine view of Forrest's motives, comparing his vision to that of Gates. "I think he is going to be one of the world's greatest philanthropists," Poynton says. Poynton notes, however, that Forrest's enemies will always point to his colourful record in business to seek to downplay his good deeds, which he believes is unfair. "Like any entrepreneur, like any leader, like any alpha character, there will always be attributes about them personally that might grate with some people," Poynton says. "But at the end of the day, you've got to look at his track record. He hasn't cashed out, he hasn't made a quick quid and moved on. He fervently believes in what he's doing, and now he's parlaying what he's made on paper into something globally significant with his philanthropy."

## 15.

## WEALTH AND POWER

*I honestly think he think he can be prime minister.*
—MARK CARUSO, Perth mining identity

Andrew Forrest tries desperately to project an image of being an ordinary, knockabout Aussie bloke whose wealth hasn't changed him. That may have been true once, but it's a much more difficult argument to mount these days.

After making his first $2 billion in 2007, Forrest said: "I still walk around barefoot with my kids on the weekend, I drive a car old enough to vote [a 1986 Mercedes], I don't gloat." A year later, when his fortune was approaching $10 billion, he continued to insist that money wasn't overly important. "Once you can look after your own needs, it becomes superfluous," he said.

But Forrest has gone to great lengths to look after those needs. In fact, he has spent tens of millions of dollars acquiring some of the trappings of wealth favoured by billionaires the world over. The beaten-up Mercedes he used to boast about is long gone from Forrest's seven-car garage in Cottesloe. In its place is a luxury Lexus hybrid, parked alongside Nicola's Lexus SUV. The Forrest home, perched at the peak of John Street near picture-perfect Cottesloe Beach, is suitably magnificent, even if slightly modest when compared with the narcissistic monuments favoured by many of Perth's other mega-wealthy residents.

Forrest has spent millions of dollars restoring the interior of the house, a sprawling dwelling built in 1895 by retail magnate Frank Zimpel. From a grand turret at the front flies the Australian flag – the only clue beyond the Norfolk Island pines lining the street that this is the domain of a patriotic mining tycoon. Those who have been inside are impressed by Forrest's underground humidified cellar and its hundreds of bottles of fine wine. There are few outward shows of pretension, however. On a normal evening, Nicola will be cooking dinner while the kids and their friends will be hanging around the kitchen or watching television – unless they're taking a dip in the heated swimming pool or playing on the floodlit tennis court.

Forrest also owns a penthouse apartment on Sydney's Circular Quay that features floor-to-ceiling windows with sweeping views of the Opera House and the Harbour Bridge on one side and the Botanic Gardens on the other. The apartment, which doubles as his Sydney office, was bought for $6 million in 2006. And he appears to enjoy the frills of owning one of Sydney's glitziest residences. When Sydney journalist Elizabeth Knight interviewed Forrest about his philanthropic work over lunch in the penthouse, they dined on quail, foie gras, grilled figs and rosella gastrique, washed down with a fine oaked chardonnay.

According to property records, Forrest also paid $1.4 million in 2007 for a property in the historic NSW village of Berrima. The splurge continued in 2010 and 2011, when Forrest bought three apartments in his own beachside suburb of Cottesloe for a total of $2.5 million. The family also has a forty-hectare retreat near Byford, in the Perth foothills, for weekend horseriding expeditions.

As his wealth rose with Fortescue's early success, Forrest began talking about upgrading his fifteen-metre boat, *Geordie Girl*, for a flash 26-metre launch that cost $10 million. He took his father, Don, to a boat show to look at the shiny new craft he coveted. Don recalled: "I said to him, 'I don't like it, you have a perfectly good boat and you don't need a bigger one.'" Forrest took the parental advice and decided to stick with his old boat.

But in late 2008, his wish for a more luxurious boat was granted when he bailed out a friend, Perth stockbroker Andrew Frazer, who had run up big losses during the global financial crisis. Forrest took a mortgage over Frazer's century-old mansion in the riverfront suburb of Peppermint Grove. He also helped out by buying his mate's 21-metre boat and beachfront property in Gracetown, which is close to the vineyards of Margaret River. At last, he had a motor launch he could proudly show off and the obligatory holiday house in the wine region of the south-west, the playground of Perth's rich and powerful.

Forrest still lacked the ultimate symbol of corporate success: a private jet. There was a time when he regarded business people who criss-crossed the continent on their own planes as self-indulgent and removed from everyday life. "He always said, 'Shoot me if I ever get a private jet,'" says one old friend. Something must have changed in Forrest's attitude. In early 2011 he arranged for Fortescue to take delivery of a $53-million Bombardier Global Express to whiz him across the nation and, increasingly, around the world. Even after he stepped down as chief executive later that year to become chairman, he continued to be a regular user of the sixteen-seat plane.

Forrest now had the Cottesloe beachside family home, the Sydney penthouse apartment, the Gracetown coastal holiday house, the Byford rural retreat, the investment properties, the luxury boat and the Fortescue corporate jet. All this for a man who claimed to eschew the accoutrements of wealth.

His good friend Rodney Adler says Forrest was once driven by the desire to become super-rich, but believes he has now moved beyond that. "When I first met him at Jacksons, all the way up to the founding of Fortescue, he was motivated by wealth," Adler says. "There's no question about that. But after you go past a certain point – and for Andrew I don't know which billion it was – the motivation is no longer wealth. Because you have so much, you can't spend it in your lifetime."

The real jewel in Forrest's empire is Minderoo, the 280,000-hectare cattle station in the Pilbara that served as the family seat for four generations until Don was forced to sell in 1998. Forrest was shattered by his father's forced sale of their ancestral land, vowing to buy it back one day if he ever had the money. Ten years later, when his bank balance was much healthier, he began calling Minderoo's owners – Kalgoorlie mining identities Peter Bartlett, Gary Connell, Graeme Smith and Ron Harken – to ask them to sell. The owners demanded $30 million for the station, which Twiggy refused to pay. But when the property was put up for auction in 2009, Forrest sent in two private bidders and bought Minderoo for $12 million. After eleven agonising years, the family name was back on the mailbox.

Forrest has since spent millions of dollars restoring the main homestead to its former glory and refurbishing more than a dozen other buildings on the property. In 2012, he invited several journalists – including the author – to fly to the Pilbara and stay at Minderoo. During a tour of the station, Forrest showed the old two-way radio he used for School of the Air lessons and, while strolling through the shearing quarters, recalled the boyhood thrill when his father and the head shearer would give him a sip from their beer bottles at the end of the day. For the cameras, he also put on a display of his jackerooing skills, helping to push a herd of Red Angus cattle through the stockyards.

That night, at the end of a sticky day in the red dust, Forrest pulled on his board shorts, walked across the homestead's superbly manicured lawns and jumped into the soothing water of the man-made lake. Later, he produced bottles of red wine from his cellar and chatted for hours. It was proof that Forrest, when he turns on the charm, is good company. Next morning, he was up early cooking fillet steak for his guests' breakfast.

When Forrest bought Minderoo, he believed he was righting a historical wrong. For his parents and siblings, it had been distressing to have the station out of the family's hands for so long. "To see the look on my brother's face, to hear the voice of my sister and to look into my father's eyes and my mother's eyes and see their joy is something I will never forget," he said of buying back the property. "The place we had put so much heart into and so much blood, sweat and tears, was now back in the family and in the hands of the people who loved it. If you work hard, Australia affords you the opportunity to literally buy back the farm."

To spend time with Forrest at Minderoo is to glimpse a man who views himself as much a pastoralist as a mining magnate or a philanthropist. He sees the station as a money-making venture, even if cattle breeding these days is a thin-margin business. After taking possession of Minderoo, Forrest employed one of Australia's best cattlemen, Phil Clark from Alice Springs, and slashed the herd from 11,000 to 3500 to ease pressure on the land.

Forrest's intense familial attachment to Minderoo may explain his bizarre attempts to keep other mining companies off the land. Like other pastoral stations in Western Australia, Minderoo is rich with minerals buried beneath it. But pastoral leases are all on Crown land, leaving pastoralists with little room to object to mining and exploration permits. In this region of wide open spaces, pastoralists and miners usually coexist quite happily.

As a miner, Forrest has taken full advantage of his rights. Fortescue's Cloudbreak and Christmas Creek mines in the Pilbara were built across three active pastoral leases, including Gina Rinehart's Mulga Downs station. Despite some objections from pastoralists in the early days about the huge scale of the project, Fortescue was able to carve its mines, railways, roads and an accommodation village that it once described as "the size of a small town" out of the pastoral land. The company went on to lock up a total of 85,000 square kilometres of prime land across the Pilbara.

But in 2010, the shoe was on the other foot for Forrest when a small private company called Yarri Mining lodged an application to conduct sand mining at Minderoo. The Minderoo lease is more than twice the size of New York City and is so vast that fully grown cattle are often brought in during a muster that have never been in the yards. Yarri applied for two mining leases covering only 141 hectares – or about 0.05 per cent of the total area. In response, Forrest's private company, Forrest & Forrest, lodged multiple objections to the grant of the mining leases, arguing in the Warden's Court that the noise from any mining operation would frighten the cattle and threaten the survival of a rare mammal called the mulgara.

In a judgment handed down in January 2013, Magistrate Stephen Wilson took only a few sentences to sum up what he thought of Forrest's central arguments. It was difficult to fathom, he said, how two mining leases covering such a small area could cause any financial or environmental impact to Minderoo. But Wilson saved his real disdain for Forrest & Forrest's "dictatorial" application to the court to impose a condition that Yarri must pay Forrest an annual performance guarantee of at least $200,000 if mining went ahead. "In my opinion this proposed condition by Forrest is outrageous," he said. "The proposed condition appears to be an attempt to create some form of dictatorial power that allows it to act as the investigator, prosecutor, judge and enforcer of the provisions of the Mining Act when it has no power or right to do so."

Later, Andrew Forrest also opposed an application by Perth mining identity Tony Sage to explore for uranium on Minderoo. Sage, who had spent years working closely with Pilbara station owners, was shocked that a man who had championed the junior end of the mining sector could object to the use of pastoral land by an explorer. "We are disappointed that someone so prominent in the Australian mining industry, and who ... has purportedly 'fought' for junior exploration companies, has taken such a bewildering stance," Sage said.

This wasn't the first time his fellow miners had accused Forrest of hypocrisy. When he created Fortescue, Forrest promised repeatedly that his railways would be open to other Pilbara miners and this would help unlock the huge potential of the region for anyone with a stranded deposit. This open-access regime was also a critical part of the mining agreement he signed with the WA government in 2004. Forrest's mantra was that BHP Billiton and Rio Tinto had failed the nation by refusing access to their infrastructure, but he would be its saviour. When he opened Fortescue's Pilbara operations in 2008, Forrest declared with typical hyperbole: "These ports and lines will have lives way beyond ours. These are the arteries which will see hundreds of other projects grow."

Years later, however, Fortescue's rail lines remained as closed to other companies as those of BHP and Rio. By 2013, it was painfully clear to the up-and-coming miners in the Pilbara that Fortescue would not allow them rail access without attempting to extract a heavy price. Fortescue did reach an agreement in 2011 to transport ore for one company, BC Iron, but only after driving a hard bargain that involved it securing a 50 per cent equity stake in BC Iron's mine.

Another junior miner, Brockman Mining, became so frustrated with Fortescue's attitude that it applied to the WA government, using an unused part of the railways access code to try to force the company to give it access to part of its 280-kilometre line. Fortescue demanded up to $576 million a year – a massive price – from Brockman to use the railway. It later said it would not have any spare capacity on its line for twenty years, and that the Chinese-owned Brockman would need to pay for an expansion of the railway if it wanted to use it. Fortescue argued in submissions to the WA Economic Regulation Authority that allowing Brockman's trains on its railway would create "material inefficiencies" in its train scheduling and maintenance. The claim echoed those made by BHP Billiton and Rio Tinto in the 2000s, when they successfully fought Fortescue's own efforts for third-party access.

And therein lie the contradictions at the heart of Forrest and the mining empire he has built in the Pilbara dirt. This is a man for whom life is a sequence of ambiguities and paradoxes. He wants to be seen as an everyman who took on powerful mining giants and triumphed for the benefit of all Australians, yet he regularly conflates the national interest with his own commercial interests. He is an elitist who craves the respect of the establishment but wants to be loved by ordinary people. An honest bloke found by several courts sometimes to have trouble with the truth. A devout Christian who does business with morally dubious characters. A philanthropist who will do anything for a dollar. A supporter of Aboriginal people except those who stand in the way of his mines.

What is next for Andrew Forrest? Still only in his early fifties, and with energy to burn, he will likely be a prominent figure in Australian public life for decades to come. Forrest has stepped down from day-to-day duties at Fortescue and handed the job to a management team that no longer needs a master promoter at the helm. He has thrown some of his money at a few promising nickel and gold mining plays, though none will ever be as spectacularly successful as Fortescue. And he has embarked on a series of philanthropic ventures that are bigger and bolder than anything else undertaken in this country.

His most obvious career move would be politics. He is close to Tony Abbott and the two men have chatted informally about

Forrest's interest in joining the Liberal Party and possibly entering federal parliament. Forrest has sounded out those close to him about whether he should go into politics. If he does put his hand up, few of his old friends and acquaintances would be surprised.

Mark Caruso, the man who sold Forrest the tenements that made him a multibillionaire, believes Forrest's overwhelming motivation in life is restoring his family name to greatness in public life, and being prime minister would be the ultimate way of achieving this. "He has both the influence and resources to become a significant politician," says Caruso. "I honestly think he can be prime minister."

Albert Wong, Forrest's buddy from his Sydney deal-making days, is another who reckons he will aim for the prime ministership. "One day, I said to him, 'You've done it all now, you might as well be prime minister.' He didn't admit it to me, but I'm sure that has crossed his mind. Ultimately, Andrew is the type of guy that really wants to make his mark and leave a legacy."

Warwick Grigor believes his old business partner is more likely to follow in the footsteps of Sir John Forrest and seek to become the premier of Western Australia. "He has a need to do what his forefathers have done," Grigor says. "I told him one day he would be premier of Western Australia and that was his overriding desire, and he looked at me and said, 'How do you know that?'"

Others who know Forrest say he is fixated by how he will be remembered in fifty years' time. "Twiggy won't be happy with a bronze statue in Kings Park [like Sir John Forrest]. He'll want a gold one," a colleague told the *Australian*.

One close friend reveals that Forrest is still trying to work out how to make a contribution to Australian public life, and this may include a career in politics. "But I think it would kill him – he would have to toe the party line," the friend says. "He would want the red carpet treatment, but he could never go straight in as prime minister, and I think he would only ever have aspirations to be prime minister."

Whatever his aspirations, Forrest will seek to differentiate himself from the fumbled efforts of his fellow mining billionaires, Gina Rinehart and Clive Palmer, in seeking to exert influence in Australian life. Only a handful of people were aware that Forrest, in 2011, rejected the major stake in publisher Fairfax Media that Rinehart eventually bought, a move that signalled he had no desire to become a media mogul. And while Palmer created his own political party to contest the 2013 federal election, it's unlikely even Forrest would see any advantage in such an ego-driven stunt.

Even if he decides against a shot at politics, Forrest is deftly positioning himself as a 21st-century statesman. At the high-level Boao Forum for Asia conference in 2013, Forrest led the Australian delegation and hosted cocktail parties in his hotel suite for government and corporate leaders from Australia and China. Of all the delegates, he was the most passionate in arguing that Australia must work harder to boost its trade and investment links with a country that he insists will remain an economic powerhouse for decades. Those present witnessed a businessman with the ability to schmooze the Chinese leadership better than anyone.

Yet there are still those who believe Forrest's enduring belief in his own infallibility will ultimately prove to be his downfall. "In the final analysis he will fall victim to his own personality, which is to keep pushing and pushing and pushing," warns Brian Burke, the former Western Australian premier who worked closely with Forrest. "One day there will be something that *will* be impossible and Twiggy won't recognise it." Burke pauses and searches for an example. "I will never run 100 metres in ten seconds," he says. "But if you told Twiggy, he'd say, 'Let me have a few months' training and I will do it.' In the final analysis, that inability to accept that something is impossible will cause him to fail."

The great unanswered question of Forrest's life is whether he is motivated more by money, power or the need for respect. The answer will, to a large extent, determine how he is remembered. Whatever happens, though, Forrest's fortunes will continue to be governed by chance. If the Chinese economy suffers a serious jolt, his wealth and influence could shrivel up almost overnight. Under this scenario, he would be left without much of a fortune to give away to charity. And if he does decide on a political career, he will have to rely on the vagaries of the party system, and the will of the people, to get to the top.

Forrest has never chosen the safe path. Driven by the legend of his pioneering ancestors, he continues to dream of conquests on a grand scale, and he is desperate to prove his remaining doubters wrong. Of this we can be sure: the high-stakes life of Andrew Forrest has plenty more adventures ahead.

# ACKNOWLEDGMENTS

I would like to thank the scores of people who helped me research this book by agreeing to speak to me about Andrew Forrest. Many are prominent in business, politics and other fields, and most spoke on the condition of anonymity for reasons that are understandable.

My boss at the *Australian*, Geoff Elliott, could not have been more supportive when I approached him with the idea of writing a book. He gave me the time to write, provided advice and tolerated my absences from the office. The editor of the *Australian*, Clive Mathieson, also encouraged the project. I am also indebted to my colleagues Milan Scepanovic and Colin Murty, who went beyond the call to help me source photographs. Tony Barrass, the best writer in the business, read the entire manuscript without a word of complaint and suggested plenty of improvements. Rebecca Turner was both a superb "in-house" editor and constant sounding board.

Others who provided advice or help along the way include Paul Downie, Paul Garvey, Glenn Burge, Peter Klinger, Paul Cleary, Peter van Onselen, Paige Taylor, Tom Baddeley, Jan Mayman, Kate Askew, Chelsea Taylor, Nic Perpitch and Michelle Kiddie.

In my research, I drew heavily on excellent personal profiles of Andrew Forrest published over the years by newspaper journalists Cameron Stewart, Mark Drummond, Ben Hills and Elizabeth Knight.

At Black Inc., Chris Feik's guidance and vision throughout the project were invaluable. The supremely patient Nikola Lusk turned my manuscript into a real book. Thanks also to Elisabeth Young, Anna Lensky and Sophy Williams.

I could not have written this book without such love and support over the years from my family, especially from my parents, Lynne and John.

Above all, I would like to thank Bec, simply for making it all possible.

# NOTES

### Prologue
"An idiot who got lucky": author interview with anonymous Perth business leader, 2012.

"the hallmark of a great entrepreneur": Daniel Goleman, "The Psyche of the Entrepreneur", *The New York Times*, 2 February 1986.

"the will to conquer": Joseph A Schumpeter, *The Theory of Economic Development: An Inquiry into Profits, Capital, Credit, Interest and the Business Cycle*, Harvard University Press, Cambridge, MA, 1934.

### Chapter 1: The Forrest Legacy
"So much of what he does is proving to his mother": author interview with Warwick Grigor, 2013.

"Andrew would like to emulate his great-great-uncle": Don Forrest to Cameron Stewart, "The Accidental Billionaire", *The Weekend Australian* magazine, 24 May 2008.

"We can never escape who we are": Andrew Forrest to Mark Drummond, "Forrest Fire", *The West Australian*, 1 September 2007.

"those two guys probably did have an impact": Andrew Forrest, interviewed by Peter Thompson, Centre for Social Impact, 27 May 2011.

"[the family legacy] is on his shoulder": author interview with Graeme Kirke, 2012.

"He's seriously got whatever it was": Janie Hicks, *Dynasties*, ABC TV, 28 November 2005.

"Andrew will probably go down as a very significant part": David Forrest, *Dynasties*.

"Twiggy has said his business motto today": Elizabeth Knight, "Lunch with Andrew Forrest", *The Sydney Morning Herald*, 12 June 2010.

"as successful Olympic athletes": Geoffrey Bolton, *Land of Vision and Mirage: Western Australia since 1826*, UWA Press, Crawley, WA, 2008, p42.

"Water became scarce and grass gave way to spinifex": Cyril Ayris, *John Forrest: Man of Legend*, Cyril Ayris, West Perth, WA, 1996, p27.

"his achievements are largely ignored": Bolton, *Land of Vision and Mirage*, p42.
"local boy made good": Frank Crowley, *Big John Forrest 1847–1918: A Founding Father of the Commonwealth of Australia*, UWA Press, Crawley, WA, 2000, p58.
"Why did he seek the premiership so ardently?": Crowley, *Big John Forrest*, p79.
"Railways, harbours, everything": Crowley, *Big John Forrest*, p101.
"waste and extravagance": Bolton, *Land of Vision and Mirage*, p63.
"promoted an ideology of development": Bob Reece & Tom Stannage (eds), *European–Aboriginal Relations in Western Australia: Studies in Western Australian History*, Vol. 8, UWA Department of History, Crawley, WA, 1984.
"until the race died out": Crowley, *Big John Forrest*, p298.
"It's all very well for us to be incensed": Crowley, *Big John Forrest*, p298.
"There are millions of them": Crowley, *Big John Forrest*, p203.
"this compound of social snobbery": Crowley, *Big John Forrest*, p5.
"used his own money to finance expeditions": Alison & Dinee Muir, *Forrest Family, Pioneers of Western Australia*, JR Muir & Son, Manjimup, WA, 1982, p85.
"played the stockmarket with zest": Muir, *Forrest Family*, p88.
"it would have been almost impossible": Muir, *Forrest Family*, p90.
"As a capitalist he was an uncomplicated believer": GC Bolton, "Forrest, Alexander (1849–1901)", *Australian Dictionary of Biography*, Volume 8, Melbourne University Publishing, South Carlton, Vic., 1981.
"He died with an estate worth £195,238": Bolton, "Forrest, Alexander".
"My dear Deakin, 'Et tu, Brute?'": Crowley, *Big John Forrest*, p423.
"I have faced death before": Muir, *Forrest Family*, p85.

Chapter 2: Minderoo
"We could get on a horse and ride" and "You made your life with the animals": Andrew Forrest and Janie Hicks, *Dynasties*.
"These kids were smarter than me": Andrew Forrest, Speech to 10th National Business Leaders Forum on Sustainable Development, May 2009.
"Mum pulled me out of the hostel": Knight, "Lunch with Andrew Forrest".
"a much more extended family": Andrew Forrest, Speech to National Business Leaders Forum.
"I was in awe of him": Billy Rule, "Dance of Destiny", *The Sunday Times*, 27 June 2010.
"you've just done a bloody man's job": Rule, "Dance of Destiny".
"the 'wild native' drove his spear": Muir, *Forrest Family*, p101.
"David told the magistrate": Muir, *Forrest Family*, p101.
"a real at-oneness with the land": Forrest, interviewed by Peter Thompson.
"My strongest memory of Grandad": *Dynasties*.

"I don't think Dad enjoyed it" and "He used to tell me what to do": *Dynasties*.
"When it was stinking hot": *Dynasties*.
"I could cope with the heat": *Dynasties*.
"I don't know what I went into": *Dynasties*.
"Don forced him to chew a mouthful of soap": Drummond, "Forrest Fire".
"Don was reluctant to tell his son he was proud of him": Stewart, "The Accidental Billionaire".
"it was his strong mother": Knight, "Lunch with Andrew Forrest".
"We came across industry of sight and sound": *Dynasties*.
"Forrest would return to the Cape Lambert project": Michael Weir, "Forrest Rises Above Doubters", *The West Australian*, 30 May 1998.
"the huge Mount Whaleback mine had engaged": Paul Cleary, *Too Much Luck: The Mining Boom and Australia's Future*, Black Inc., Collingwood, Vic, 2011, p132.
"With little demand for their skills": Bolton, *Land of Vision and Mirage*, p157.
"he believed 'no-good half-caste'": Lang Hancock, quoted in the documentary *Couldn't Be Fairer*, Dennis O'Rourke (producer), 1984.
"It was always hateful, absolutely dreadful": *Dynasties*.
"you'd try and get as much red dust": *Dynasties*.

## Chapter 3: Fisticuffs and Failures

"I always had the stutter": Forrest, interviewed by Peter Thompson.
"I'm amazed he speaks so well now": Drummond, "Forrest Fire".
"likening the environment to the savagery": author interview with Tony London, 2012.
"Moyes was 'well past his use-by date'": author interview with anonymous former Christ Church student, 2012.
"Forrest taunting a fellow student": author interview with anonymous former Christ Church student, 2012.
"boomerang elbow": Drummond, "Forrest Fire".
"He was very quick": author interview with anonymous former Christ Church student, 2012.
"it was lucky he had a sheep and cattle station": Drummond, "Forrest Fire".
"He was a very unhappy boy" and "Andrew divided people": author interview with Tony London, 2012.
"He was fighting one morning": author interview with anonymous former Christ Church student, 2012.
"I was standing at the top of the stairs" and "After our encounter": author interview with John Weatherhead, 2012.
"a joke about Forrest's stutter": author interview with Bruce Strapp, 2012.
"I didn't want to pry": author interview with Ken Tregonning, 2013.
"There was no way if he stayed at Christ Church": author interview with Rohan Pixley, 2012.

"I used to relish boys who were in trouble" and "He told me long after": author interview with Ken Tregonning, 2013.
"he'd end up pushing the envelope a bit": author interview with anonymous former Hale student, 2013.
"we realised he was a snake oil salesman": Forrest, interviewed by Peter Thompson.
"It was a real blood sport": Knight, "Lunch with Andrew Forrest".
"His stutter was still quite pronounced" and "He was very persistent in asking girls out": author interview with anonymous friend of Andrew Forrest, 2013.
"the fact that Forrest repeated Year 11": author interview with Ken Tregonning, 2013.
"He was reasonably good academically": author interview with anonymous former Hale student, 2013.
"The world is full of intellectuals that failed": Knight, "Lunch with Andrew Forrest".
"when boarders from yesteryear make good": author interview with Bill Edgar, 2013.
"a huge wave sank the small fishing boat": Stewart, "The Accidental Billionaire".
"I'd kind of been given the impression": Forrest, interviewed by Peter Thompson.
"He was never really interested in cattle and sheep": Stewart, "The Accidental Billionaire".

## Chapter 4: The Wild West

"It was hard to be serious": author interview with John Poynton, 2013.
"terribly enthusiastic young man": Drummond, "Forrest Fire".
"People liked going to Laurie for money": Anne Lampe & Colleen Ryan, "How Laurie Sank a Bank", *The Sydney Morning Herald*, 28 January 1989.
"Connell was siphoning off more than $130 million": Trevor Sykes, *The Bold Riders: Behind Australia's Corporate Collapses*, Allen & Unwin, Crows Nest, NSW, p96.
"Depositors [of Rothwells] would surely have been alarmed": Sykes, *The Bold Riders*, p96.
"He was always on two phones", "We had to buy him a suit" and "We all went to the window": author interview with Graeme Kirke, 2012.
"the 'incredibly brave' Forrest jumped in": Paul Lampathakis, "'Brave' Twiggy Forrest Punched as He Breaks Up Street Fight", *The Sunday Times*, 4 December 2010.
"between Laurie and a bag of money": Sykes, *The Bold Riders*, p70.
"I don't think Andrew could ever work without": author interview with Graeme Kirke, 2012.
"That was something adults did": author interview with anonymous former colleague of Andrew Forrest, 2012.

"Jacksons were go-getting blokes": author interview with Peter Richard, 2013.

"got his break in the early 1980s": Mark Drummond, "The '80s Upstart Who's Come Home to Spend his Fortune", *The West Australian*, 5 May 2007.

"Andrew idolised Dave": author interview with anonymous former colleague of Andrew Forrest, 2012.

"he never quite stayed out of trouble": "Trading in Max Poised to Resume as Probe Concludes", *Financial Times*, 27 October 2007.

"I wasn't sure whether to go": author interview with Jeff Braysich, 2012.

"many of the older brokers along St Georges Terrace": author interview with John Poynton, 2013.

"We weren't exactly loved", "They did their big deals" and "Andrew comes straight in": author interview with Jeff Braysich, 2012.

"We didn't know any better" and "You were either his best mate": author interview with Jeff Braysich, 2012.

"He was a mentor" and "He was happy to go out": author interview with anonymous former colleague of Andrew Forrest, 2012.

"There are some excellent situations": "Shell-Shocked Brokers Remain Nervous", *The Australian Financial Review*, 22 October 1987.

Chapter 5: Sin City

"It was Laurie Connell" and "The normal procedures you'd follow": author interview with Peter Richard, 2013.

"remembers Forrest as a skilled raconteur": author interview with Ross Dobinson, 2013.

"A lot of people criticise him": author interview with Ross Dobinson, 2013.

"he outlined plans for Intersuisse": Elizabeth Knight, "Forrest Resigns from Jacksons", *The Sydney Morning Herald*, 6 May 1988.

"Forrest made at least forty good friends": author interview with David Hannon, 2013.

"I did not want to, nor could I": author interview with Rodney Adler, 2013.

"Andrew with his silver tongue", "Andrew taught me how to fly first-class", "Any normal, sane person wouldn't have done it" and "I'm going to take Nicola out": author interview with Albert Wong, 2012.

"He had a real weakness for the ladies": author interview with anonymous friend of Andrew Forrest, 2013.

"He got a bit tired of talk of bridal dresses" and "Nic has been very good for him": Stewart, "The Accidental Billionaire".

"all the skeletons started to come out of the closet" and "Andrew's natural instinct": author interview with Albert Wong, 2012.

"promising investors that the alpacas would deliver": "Alpaca as Investment", *The Australian Financial Review*, 16 June 1992.

"demonstrated that he would conduct his business": International Alpaca Management & Ors v Ensor & Ors [1995] FCA 1990 (7 April 1995).

"self-assured and at times amused by his own evidence": International Alpaca Management & Ors v Ensor & Ors [1995] FCA 1990 (7 April 1995).
"Anyone who knows me will tell you": Ben Hills, "How to Win Friends and Infuriate Anglo", *The Sydney Morning Herald*, 26 May 2001.
"the business venture was 'quite marginal'": author interview with Rodney Adler, 2013.
"Everyone advised me not to do it" and "In the first three years I brought in": author interview with Warwick Grigor, 2013.
"Albert, you know the future is Asia": author interview with Albert Wong, 2012.
"most substantially financed, racist organisation": Human Rights and Equal Opportunity Commission, *Report of the National Inquiry into Racist Violence in Australia*, 1991.
"A front group created to contest": Ramon Glazov, "Prince of the Pilbara", *The Monthly*, July 2013.
"for the Christian faith to command": *The New Times*, newsletter, Australian League of Rights, January 1992.
"probably wasn't a good idea": author interview with Warwick Grigor, 2013.
"The League represents the respectable face of racism": Human Rights and Equal Opportunity Commission, *Report of the National Inquiry into Racist Violence in Australia*, 1991.
"Jeremy Lee, was also a key figure": Ramon Glazov, "Prince of the Pilbara", *The Monthly*, July 2013.
"share a commitment to Christianity": author interview with David Thompson, 2013.
"he believed his new digs would prove to be a sound investment": Kathryn Bice, "Hewson sells house for $1m", *The Australian Financial Review*, 24 May 1993.
"He had a fairly cavalier approach", "Andrew was a little bit out of control" and "Andrew had a favourite term": author interview with Warwick Grigor, 2013.

CHAPTER 6: DREAMS IN THE DESERT
"Salter returned to business only to be sacked": Andrew Burrell, "Burnt Investors Get at Payne's to Explain", *The Weekend Australian*, 3 December 2011.
"He came to me and said": author interview with Rodney Adler, 2013.
"probably feels a bit constrained": Damon Frith, "Magnesium Next for Anaconda", *The Australian*, 29 July 1999.
"Everyone who's done business with me": Damon Kitney, "From Despair to a Dream Fulfilled in a Year", *The Australian Financial Review*, 3 November 1997.
"He also boasted of how Anaconda": Tom Baddeley, "Court Criticism a New Blow to Nickel Project, *The West Australian*, 25 March 1997.

"he'd clearly had a good lunch": Mark Drummond, "Racy Russet", *The West Australian*, 29 March 1997.

"He said …. I was pathetic": Baddeley, "Court Criticism a New Blow".

"Forrest, who wasn't represented by a lawyer in court": Hills, "How to Lose Friends".

"If someone else could have achieved": Michael Weir, "I Would Trade Fee for Time: Forrest", *The West Australian*, 18 November 1997.

"That was one of the toughest times": "The Mine Field", *Four Corners*, ABC TV, 12 August 2002.

"some companies have trouble financing a new car park": John McIlwraith, "Skill at Raising Money Baffles Rivals", *The Australian*, 2 August 1999.

"He later admitted to bribing officials": Marc Rich, "'King of Oil' Pardoned by Clinton, Dies at 78", *Reuters*, 26 June 2013.

"They were all harebrained ideas": author interview with anonymous former Anaconda executive "A", 2013.

"As a CEO there's nobody": author interview with anonymous former Anaconda executive "B", 2012.

"We argued with Glencore": Andrew Forrest, interview with Ticky Fullerton, "The Mine Field", *Four Corners*.

"The autoclaves are performing": Tim Treadgold, "Crunch Time for Nickel Hopefuls", *BRW*, 29 January 1996.

"although the flash vessels were not working": Barry Fitzgerald, "Anaconda Shares Fall into Murrin Acid Bath, *The Age*, 17 March 1999.

"Forrest would admit that the plant was seriously flawed": Andrew Forrest, interview with Ticky Fullerton, "The Mine Field", *Four Corners*.

"I've met a lot of managing directors": Matt Price, "Caught Out in a Dangerous Game", *The Australian*, 5 March 2007.

"Andrew was ringing me up": author interview with Wayne Richards, 2012.

"the 'nightmares' of the previous two years": Andrew Forrest, speech at Murrin Murrin opening, 29 July 1999.

"Where companies and individuals strive to achieve": Forrest, speech at Murrin Murrin opening.

"With Twiggy, you're either totally for him": author interview with Paul Armstrong, 2012.

"I'm not interested in games": Edensor Nominees v Anaconda Nickel [2001] VSC 502 (18 December 2001).

"There has not been an obligation": Andrew Main, "Gutnick on Defensive as Centaur Breaks Leg", *The Australian Financial Review*, 11 December 2000.

"pursued a 'deliberate plan'" and other findings: Edensor Nominees v Anaconda Nickel.

"Gutnick was a truthful witness": Edensor Nominees v Anaconda Nickel.

"It's a strange judgment": Mark Drummond, "'Untruthful' Forrest Hits Out at $10m Judgment", *The West Australian*, 20 December 2001.

"very clever Queen's counsel": Forrest, interview with Ticky Fullerton, "The Mine Field", *Four Corners*.

"Anyone who knows me": Drummond, "'Untruthful' Forrest Hits Out".

"I prefer the evidence of Mr Spohler": Van der Sluys and anor v Anaconda Nickel Nl and others [2002] NSWSC 673 (31 July 2002).

"both truthful and substantially accurate": Anaconda Nickel v Tarmoola Australia, unreported, Supreme Court of Western Australia (24 February 1999).

"The numbers were tight": Sharon Kemp, "Ralph Recalls Power of One", *The West Australian*, 24 July 1991.

"the idea had probably germinated": Albert Wong, quoted in "The Mine Field", *Four Corners*.

"missed targets, lost production": Hills, "How to Lose Friends", *The Sydney Morning Herald*, 26 May 2001.

"The winner is an independent Anaconda Nickel": "Truce at Anaconda, Forrest to Surrender Chief's Role", *The West Australian*, 1 June 2001.

"Andrew knew it all": author interview with anonymous former Anaconda executive "D", 2013.

"It taught me cynically": Andrew Forrest, quoted in "Never, Ever Give up – The Fortescue Metals Group Story", video, December 2012.

"There's no doubt it wouldn't have happened": author interview with Peter Johnston, 2013.

"I remember Andrew coming through the office": author interview with anonymous former Anaconda executive "E", 2012.

"financial pressure": Forrest v Commissioner of Taxation [2010] FCAFC 6 (5 February 2010).

"narcotics-related exercise": Malcolm Brown, "Huge Kings Cross Heroin Racket Busted, Court Told", *The Sydney Morning Herald*, 27 August 1987.

"they would receive returns of 30 per cent": "Equity-1 Ltd: investor warning", ASIC media release, 30 January 2003.

"he believes he and Twiggy bonded that night", "I called the guy at the bank" and "Forrest told me": author interview with Shayne Heffernan, 2013.

### Chapter 7: The Comeback Kid

"To create an Australian-owned": Andrew Forrest, presentation to Allied Mining and Processing shareholders, 18 July 2003.

"I thought, [it] will be like Napoleon said": Drummond, "Forrest Fire".

"Andrew said, 'Come and have a scotch'": Paul Lampathakis, "Man of Steel", *The Sunday Times*, 18 May 2008.

"I laid out a map of the Pilbara": Forrest, quoted in "Never, Ever Give Up".

"After that first meeting with Andrew": Graeme Rowley, quoted in *A Sense of Place – The Fortescue Story*, Fortescue Metals Group, 2008.

"A few friends of mine were very questioning": Chris Catlow, quoted in "Never, Ever Give Up".

"That first meeting with Andrew reminded me": email to author from John Hancock, 2013.

"corporate cruelty": Knight, "Lunch with Andrew Forrest".

"the bravest thing he's done": Nicola Forrest, quoted in "Never, Ever Give Up".

"I knew he could sell ice to the Eskimos": author interview with Mark Caruso, 2012.

"remarkably selfless corporate act": Robin Bromby, "Fortescue Chief Makes a Killing", *The Australian*, 17 February 2005.

"Fortescue was started by a few friends": Andrew Forrest, address to the National Press Club, 2 May 2012.

"Andrew hugs me when he sees me", "Iron ore back then was sewerage shit", "Negotiations were colourful" and "The bank wouldn't transfer": author interview with Mark Caruso, 2012.

"It felt like the movie *Wall Street*": author interview with Philip Kirchlechner, 2012.

"It was the most disorganised show" and "We had to think outside the box": author interview with David Mendelawitz, 2013.

"We assayed the samples": Andrew Forrest, quoted in "Never, Ever Give Up".

"Andrew liked Eamon": author interview with anonymous former Fortescue executive "A", 2012.

"Morale was good": author interview with David Mendelawitz, 2013.

"I wouldn't have anyone up there": Eamon Hannon, quoted in "Never, Ever Give Up".

"Scour the entire Pilbara": Andrew Forrest, quoted in *A Sense of Place*.

"compare the work of his ancestors": Nick Evans, "Forrest Hails Hamersley Homecoming", *The West Australian*, 23 December 2012.

"it appeared to be operating in real estate": Jonathan Barrett, "Twiggy's Land Grab", *The Australian Financial Review*, 17 March 2012.

"The rejection I received was so blunt": Paul Garvey, "Critics Change Tune on Twiggy Growth Story", *The Australian Financial Review*, 31 August 2007.

"I said to [Argus] that we can both save": Herb Elliott, quoted in "Never, Ever Give Up".

"cutting back on its exploration work": Peter Thompson and Robert Macklin, *Big Fella: The Rise and Rise of BHP Billiton*, William Heinemann, North Sydney, NSW, 2009, p424.

"You hear about new ideas every day": Christopher Webb, "Chip Lets Fly at Ore-Struck Promoters", *The Age*, 17 February 2005.

"lodged challenges over more than 100 tenements": John Phaceas, "Anaconda in Spoiling Play", *The Australian*, 3 December 2001.

"BHP dumped its long-standing broker": Mark Drummond, "BHP Dumps JP Morgan over Link to Fortescue", *The West Australian*, 25 August 2007.

"most analysts will only cover a stock": author interview with Greg Lilleyman, 2012.

"when a stock rises on rhetoric and promotion": Adele Ferguson, "Iron Man", *BRW*, 4 August 2005.
"The more they threw at us": John Clout, quoted in "Never, Ever Give Up".
"If they'd embraced us a little more": Mark Thomas, quoted in ""Never, Ever Give Up".

CHAPTER 8: BURKE'S BACKYARD
"smarter than two-thirds": *The World Today*, ABC radio, 5 March 2007.
"'improper' and 'reprehensible'": Royal Commission into Commercial Activities of Government and Other Matters, report, volume 6, 1992.
"we would relish another ban": Andrew Burrell, "The Man Who Bought a Political Party", *The Australian Financial Review*, 2 March 2007.
"I told him it's extremely complex" and "BHP had contacts in the department": author interview with Brian Burke, 2013.
"Rio believed that the government approvals" and "Appropriate support should be given to companies": John Phaceas, "State, Miners at Odds on FMG", *The West Australian*, 11 November 2004.
"A state government agreement Act": *Hansard*, WA Legislative Assembly, 26 November 2004.
"the proponent has an ability to raise funds" and "The leader of the Opposition needs": *Hansard*, 26 November 2004.
"All of a sudden it gave them": author interview with Brian Burke, 2013.
"Andrew got really close to us": author interview with Julian Grill, 2013.
"They had their own people" and "I once made the mistake": author interview with Julian Tapp, 2013.
"who donated $2500 to Bowler's re-election campaign": Corruption and Crime Commission (CCC), *Report on the Investigation of Alleged Public Sector Misconduct in Connection with the Activities of Lobbyists and Other Persons – Fortescue Metals Group*, 14 September 2009.
"FMG Friday crisis" and "This was a multibillion-dollar project": CCC transcript, 27 February 2007.
"The Woodstock Abydos experience": State of the Environment Committee, *Independent Report to the Australian Government Minister for Sustainability, Environment, Water, Population and Communities*, 2011.
"The project will be killed by indecision": CCC transcript, 27 February 2007.
"We'd have had to go around" and "Minister, each time they go to blast": Andrew Forrest, quoted in "Never, Ever Give Up".
"conversation was being covertly recorded by CCC investigators": CCC, *Report on the investigation of alleged public sector misconduct*.
"As far as any lobbying goes": "Burke an excellent trainer – FMG", *AAP*, 16 November 2006.
"We did retain Burke and Grill": Marsha Jacobs, "$2 Billion Man Seizes the Moment", *The Australian Financial Review*, 19 March 2007.

"Grill also now confirms" and "Most clients took the time to express gratitude": author interview with Julian Grill, 2013.

"I don't think anyone else but Twiggy Forrest": author interview with Brian Burke, 2013.

CHAPTER 9: TO GET RICH IS GLORIOUS

"Working for a junior, it's terrible": author interview with Philip Kirchlechner, 2012.

"Every time we would go to this mill", "You had three players supplying" and "When someone else came along": author interview with Russell Scrimshaw, 2013.

"David Liu is the guy who … personifies courage": Andrew Forrest, quoted in "Never, Ever Give Up".

"laid fraud charges against Falcone": Securities and Exchange Commission, "Philip A Falcone and Harbinger charged with securities fraud", media release, 27 June 2012.

"a hyperactive, smart, energetic Australian entrepreneur": Sonali Paul, "Twiggy's Grand Bet on Australian Iron Ore Turns Sour", *Reuters*, 16 September 2012,

"I went to bed at 1am thinking": author interview with Russell Scrimshaw, 2013.

"There was a silence that seemed like forever": Russell Scrimshaw, quoted in "Never, Ever Give Up".

"greatest visionary in the Australian resources sector": Paul Garvey, "Critics Change Tune on Twiggy Growth Story", *The Australian Financial Review*, 31 August 2007.

"How he achieves things is questionable": author interview with Eddie Rigg, 2013.

"All of the project management things we'd learned": Peter Meurs, quoted in "Never, Ever Give Up".

"Everyone here is part of our Fortescue family": Jennifer Hewett, "Fortescue's Pilbara Love Story", *The Australian Financial Review*, 2 May 2012.

"Because he's a man of action": author interview with anonymous former Fortescue executive "B", 2013.

"One thing I found annoying": author interview with anonymous former Fortescue executive "B", 2013.

"If I was a little bit hesitant": author interview with Philip Kirchlechner, 2012.

"He always thought he was right": author interview with Philip Kirchlechner, 2012.

"a dispute over options": Australian Securities and Investments Commission v Fortescue Metals Group Ltd [No 5] [2009] FCA 1586 (23 December 2009).

"Their desk would be suddenly empty": author interview with anonymous former Fortescue executive "C", 2012.

"That will bring you undone": Mark Mentiplay, "FMG Fired by Vision and Empowerment", *WA Business News*, 16 February 2006.

"Everyone told us we would never succeed": author interview with David Mendelawitz, 2013.

"Nothing short of a miracle": Russell Scrimshaw, quoted in "Never, Ever Give Up".

"It's the end of five years' hard work": Andrew Burrell, "At Last, the Ship Comes in for Forrest", *The Australian Financial Review*, 30 May 2008.

"It was Andrew, basically": Paul Lampathakis, "Why Is Louise Happy?" *The Sunday Times*, 18 May 2008.

"Australian institutions accounted for a meagre 3.9 per cent": Jennifer Hewett, "Aussie Investors Too Late in Boarding Twiggy's Pilbara Gravy Train", *The Weekend Australian*, 1 December 2007.

"The company is building the most significant": Drummond, "Forrest Fire".

"I'm happy to be chained to the tracks": Matt Chambers and Damon Kitney, "Fund Manager's Railway Jibe Spurred Fortescue Founder On", *The Australian*, 2 June 2011.

## Chapter 10: The Great Escape

"If you think everything is crook in Tallarook": Michael Vaughan, "Between the Rocks and the Hard Places, *The Australian Financial Review*, 11 March 2009.

"I'm not such a great manager": Mark Ludlow, "I'm Not a Good Manager; Forrest", *The Australian Financial Review*, 21 June 2011.

"Forrest only chose that number": author interview with anonymous former Fortescue executive "C", 2012.

"Right now the focus needs to be": Jamie Freed, "Fortescue Unveils Aggressive Expansion", *The Australian Financial Review*, 12 August 2010.

"his 'severe exaggeration'": Ian Cumming and Joseph Steinberg, letter to Leucadia shareholders, 25 June 2013.

"His personality dominated the FMG board": Cumming and Steinberg, letter to Leucadia shareholders, 25 June 2013.

"I think it's a great investment": Sarah-Jane Tasker, "Forrest Puts Money Where Mouth Is", *The Australian*, 30 August 2012.

"Fortescue was in 'great shape'": Tasker, "Forrest Puts Money Where Mouth Is".

"He [Forrest] builds iron ore mines on a vast scale": Paul, "Twiggy's Grand Bet".

"Workers munched on sandwiches": Jamie Freed, "Fortescue's Near-Death Experience", *The Australian Financial Review BOSS* magazine, 10 May 2013.

"I couldn't even say goodbye": Rania Spooner, "Fortescue Tumult Continues as Secretary Resigns", *The Age*, 6 September 2012.

"Of all our assets, the ones we value most": *A Sense of Place*, p110.

"It was a very sad day for Fortescue": author interview with anonymous former

Fortescue executive "D", 2013.

"I can't tell you how well I am sleeping at night": Matthew Stevens, "Twiggy Comes Out Swinging", *The Australian Financial Review*, 8 September 2012.

"barbecues would be axed": Stephen Bell, "Barbecues Get Chopped as Miner Fortescue Cuts the Fat", *The Wall Street Journal*, 25 September 2012.

"The main story on the front page": Nick Evans, "Fortescue Sheds $1.5b as Debt Crisis Bites", *The West Australian*, 14 September 2012.

"The disaster unfolding at Fortescue": Adam Schwab, "Fortescue: The Prophet and the Loss", *Crikey*, 18 September 2012.

"describing it as 'succulent'": Leucadia National Corporation, annual report, 2011.

"Fortescue was willing to pay": Stephen Shore, "Fortescue's Great Escape", *The Weekend Australian Financial Review*, 19–20 January 2013.

"He warned that Fortescue": Stephen Shore, "Fortescue's Great Escape".

"Andrew Forrest is financially secure once more": John Durie, "Has Forrest Learned Debt Lesson?" *The Australian*, 18 September 2012.

"reports of my death": Stevens, "Twiggy Comes Out Swinging".

"It is safe to say that enterprise": John Maynard Keynes, *The General Theory of Employment, Interest and Money*, Macmillan for the Royal Economic Society, London, 1936.

"Forrest is hugely exposed to a single market": author interview with Philip Kirchlechner, 2013.

Chapter 11: China Crisis

"He announced he was seeking a partnership": Mark Drummond, "Forrest Seeks Hope Downs Partnership", *The Australian Financial Review*, 19 July 2003.

"not something we are spending a lot of time": James Chessell, "ConsMin Lends Ear to Forrest", *The West Australian*, 21 July 2003.

"a move that would boost Fortescue's chances": Chessell, "ConsMin Lends Ear to Forrest".

"Graeme was stupid to buy shares": author interview with anonymous former Fortescue executive "E", 2013.

"It was not Fortescue's intention": "Fortescue Metals Group", ASX statement, 27 April 2005.

"called on Forrest to resign": Matthew Stevens, "Forrest Must Get the Chop", *The Australian*, 3 March 2006.

"the ban would likely last": Andrew Main, "Fortescue Stirs High Court's Juices", *The Australian*, 3 October 2011.

"You talk about a $1.85 billion project": Elizabeth Sexton, "The Gilded Lily That Might Just Break a Forrest", *The Sydney Morning Herald*, 21 February 2011.

"These commitments by Chinese interests": media release, Fortescue Metals Group, 5 November 2004.

"Andrew Forrest has pulled off": Robin Bromby, "Forrest Pulls Off $1.8bn China Coup", *The Weekend Australian*, 6 November 2004.

"We have no money": witness statement of Wei Fisher, Federal Court of Australia, 21 April 2008.

"This is a very important project": witness statement of Wei Fisher.

"when Andrew Forrest saw this fax": witness statement of Wei Fisher.

"After lunch I was called" and "This is Graeme's decision": witness statement of Wei Fisher.

"Chinese groups would not pursue the project": Stephen Wyatt, "Doubts Hit Fortescue's Iron Ore Project", *The Australian Financial Review*, 24 March 2005.

"The ability of Mr Forrest": Ian Howarth, "They've Twigged to Smooth-Talking Andrew Forrest", *The Australian Financial Review*, 24 March 2005.

"Nobody knew anything about it": Sean Cowan, "Elliott Got $2m from Share Sales", *The West Australian*, 4 March 2006.

"a negotiating ploy": Nicolas Perpitch, "Chinese 'Lesson' on Fortescue Stake", *The Australian*, 9 April 2009.

"the conviction was unrelated": John Garnaut, "Corrupt Middleman Now in Beijing Jail", *The Sydney Morning Herald*, 10 April 2009.

"No investor had complained": Julie-Anne Sprague, "ASIC Case Artificial: Myers", *The Australian Financial Review*, 7 May 2009.

"ASIC's case is an attack on my client": Chalpat Sonti, "ASIC dispute 'a lot of hot air'", *The Age*, 7 May 2009.

"Forrest had not breached his duties": Julie-Anne Sprague, "ASIC Case Artificial".

"blunt commercial tactic": Australian Securities and Investments Commission v Fortescue Metals Group Ltd [No 5] [2009] FCA 1586 (23 December 2009).

"China's aim was to drive down": Andrew Burrell, "Chinese Still Keen on Iron Ore Stake", *The Australian Financial Review*, 1 May 2008.

"I have concluded that FMG's and Forrest's opinion": Australian Securities and Investments Commission v Fortescue Metals Group Ltd [No 5] [2009] FCA 1586 (23 December 2009).

"I'd like to take this opportunity to thank God": Peter Klinger & Peter Kerr, "Forrest Beats Dishonesty Charge after Long Battle", *The West Australian*, 24 December 2009.

"We say the question is": Elizabeth Sexton, "High Stakes: The Case against David Who Turned into a Goliath", *The Sydney Morning Herald*, 3 December 2010.

"whether the prosecution of this case by ASIC": Australian Securities and Investments Commission v Fortescue Metals Group Ltd [2011] FCAFC 19 (18 February 2011).

"More likely than not, many traders lost": Australian Securities and Investments Commission v Fortescue Metals Group Ltd [2011] FCAFC 19 (18 February 2011).

"mean and vengeful" and "It's on the record": Rachel Donkin, "Fortescue's Chief Claims ASIC Out for Revenge", *The West Australian*, 22 February 2011.

"Let's just say people are looking at this decision": Andrew Burrell, "Leader of the FMG 'family' Unbowed", *The Australian*, 26 February 2011.

"When you look at Fortescue": "Forrest to Appeal Federal Court Decision", *AAP*, 21 February 2011.

"I said, 'We won, Andrew'": Peter Huston, quoted in "Never, Ever Give Up".

"Fortescue's remarks were not directed": Forrest v Australian Securities and Investments Commission [2012] HCA 39, 2 October 2012.

"everyone who is contemplating an investment": Andrew Burrell, "Fortescue Case 'Hurts Investors'", *The Australian*, 4 October 2012.

"If remarks are made to the ASX": Patrick Durkin, Hannah Low & Ayesha de Kretser, "China Hype: High Court Backs Forrest", *The Australian Financial Review*, 2 October 2012.

"The majority decision shows some keen commercial insight": Durkin, Low & de Kretser, "China Hype".

"Compliance with continuous disclosure": ASIC, media release, 2 October 2012.

"Everyone saw straight through those Chinese": Forrest, quoted in "Never, Ever Give Up".

"create an environment where globally": Andrew Forrest, doorstop interview, 14 November 2012.

"just another Australian concerned": Elizabeth Knight, "Forrest Grabs a New Cause", *The Sydney Morning Herald*, 6 October 2012.

"Andrew gilds the lily": author interview with anonymous friend of Andrew Forrest, 2012.

"Seventy per cent of what he says": author interview with anonymous former Fortescue executive "D", 2013.

"He's such a good salesman": author interview with David Mendelawitz, 2013.

"I know what a lot of people have said": author interview with Rodney Adler, 2013.

"When you start with nothing": author interview with Rodney Adler, 2013.

## Chapter 12: Taxing Times

Much of the information in this chapter is taken from the author's off-the-record interviews with key participants in the mining tax debate.

"I really don't enjoy the spotlight" and "I don't want to look as if I am out there": Stewart, "The Accidental Billionaire".

"Swan had not delivered": Maxine McKew, *Tales from the Political Trenches*, Melbourne University Publishing, Carlton, Vic, pp180-181.

"The combined expertise of Gray": McKew, *Tales from the Political Trenches*, p182.

"I rang the treasurer three times": Paul Kelly, "Rescue Arrived Too Late", *The Australian*, 18 April 2012.

"It showed we would have a negative cash flow": Kelly, "Rescue Arrived Too Late".

"I'm out there employing thousands of people": *Inside Business*, ABC TV, 9 June 2010.

"paying a dollar more in tax": Joe Spagnolo, "PM Hangs Up Hotline", *The Sunday Times*, 6 June 2010.

"I'm very happy to pay tax": "Fortescue Chief Sees Deal on Mining Tax", *AAP*, 27 June 2010.

"John Forrest was determined to ensure": Robert Taylor, "Billionaires, Workers Rail against New Tax", *The West Australian*, 10 June 2010.

"I have a prime minister who is listening": David Crowe, "PM's Listening: Forrest", *The Australian Financial Review*, 11 June 2010.

"FMG would pay very little": Treasury briefing by Scott Bartley, 23 June 2010.

"while a headline rate of 40 per cent": Kelly, "Rescue Arrived Too Late".

"Fortescue would not sign any separate agreement": Jennifer Hewett, "Opportunity Lost to Build Australia", *The Australian Financial Review*, 18 April 2012.

"all of the companies were not at the table": Andrew Burrell, "Miners Angry at Forrest 'Rudd Pact'", *The Australian*, 1 July 2010.

"What is good for FMG": Burrell, "Miners Angry at Forrest 'Rudd Pact'".

"We didn't realise that BHP and Rio": Kelly, "Rescue Arrived Too Late".

"Fortescue would 'carry the burden'": "Fortescue Dismisses MRRT Draft Legislation", media release, 13 June 2011.

"It was a really gruelling time for Andrew": author interview with anonymous friend of Andrew Forrest, 2013.

"irresponsible economic management": Andrew Burrell, "Resources Tax Could Be Buried by Result", *The Australian*, 23 August 2010.

"My heart doesn't break for Fortescue": Jennifer Hewett, "Twiggy Delivers Rebuke to Kloppers", *The Australian*, 22 September 2010.

"threat to democracy": Wayne Swan, "The 0.01 Per Cent: The Rising Influence of Vested Interests in Australia", *The Monthly*, March 2012.

"I fear Australia's extraordinary success": Swan, "The 0.01 Per Cent".

"The average punter only gets to vote": *Lateline*, ABC TV, 5 March 2012.

"I hardly think that Twiggy Forrest": Patrick Durkin & Michael Smith, "Swan 'playing man, rather than ball'", *The Australian Financial Review*, 6 March 2012.

"It is Swan's job": Malcolm Turnbull, "Not Classy, Wayne", *The Australian Financial Review*, 16 March 2012.

"All it takes for evil to conquer": "Mining Tax Evil, Says Fortescue Founder", ABC radio, 16 November 2012.

"this case should be seen for what it is": Commonwealth submission, Fortescue Metals Group & Ors v The Commonwealth of Australia, High Court, 2012.

"I think Andrew Forrest is owed an apology": *The World Today*, ABC radio, 13 February 2013.

CHAPTER 13: THE INTERVENTION

"the town's notorious jail housed": WA Department of Indigenous Affairs, *Roebourne Report: Issues, Current Reponses and Strategies for Consideration*, Government of Western Australia, July 2009.

"remarkably strong sense of connection": Daniel v Western Australia [2003] FCA 666 (3 July 2003).

"My grandfather wanted me to realise": author interview with Michael Woodley, 2013.

"before welfare 'destroyed' them": Andrew Forrest, address to the National Press Club, 2 May 2012.

"Suddenly, you don't have an economic model": Joel Gibson, "A Way Out of the Wilderness", *The Age*, 4 October 2008.

"'welfare lobby' that is 'powerful' and 'nasty'": Forrest, interviewed by Peter Thompson.

"idiot, feelgood white Australia": Victoria Laurie, "Twiggy's Intervention", *The Australian*, 20 February 2008.

"When you hear 'smoke for a poke'": Forrest, interviewed by Peter Thompson.

"They bought the most powerful": Forrest, interviewed by Peter Thompson.

"Ian Black was the prized son": Andrew Forrest, interviewed by Peter Thompson.

"Working makes me feel like I've got structure": GenerationOne video, 2012.

"We don't want to be left behind": Kris Dixon, quoted in "Never, Ever Give Up".

"We want this contract": Fortescue, ASX announcement, "Fortescue Awards More Than $1 Billion in Contracts to Aboriginal Businesses", 6 August 2013.

"the disparity in living conditions": Marcia Langton, "Busted by the Boom", *The Weekend Australian*, 24 April 2010.

"Whenever an Aboriginal group negotiates": Marcia Langton, 2012 Boyer Lectures, 18 November 2012.

"they are being lured into high-paid mining jobs": author interview with Michael Woodley, 2013.

"They think we will piss it up the wall": author interview with Michael Woodley, 2013.

"Yarri's operations would affect the viability": Nick Evans, "Forrest Loses Latest Round", *The West Australian*, 31 January 2013.

"we have always been one Yindjibarndi people": Steve Pennells, "Red Earth Dreaming, *The Weekend West*, 23–24 April 2011.

"it would not 'pay quite the same money'": "Iron and Dust", *Four Corners*, ABC TV, 18 July 2011.

"Fortescue had 'created' the splinter group": "Conflicting Information, Interests and Land Owners Dog FMG", *7.30*, ABC TV, 20 November 2012.

"On the face of it, this was an Aboriginal meeting": "Conflicting information, interests and land owners dog FMG", *7.30*.
"it did incite racial hatred": Pennells, "Red Earth Dreaming".
"more than 20,000 times": "FMG Great Native Title Swindle Part 1", video, https://www.youtube.com/watch?v=6w_fB7e0WCY.
"been viewed about 8000 times": "The True Yindjibarndi Story", video, http://www.youtube.com/watch?v=VGZ_H_SWiQM.
"It's the same old story": Stephen Hagan, "A Flawed Native Title Meeting That Could Be Twiggy's Curse", *National Indigenous Times*, 14 April 2011.
"It was the worst, most reprehensible experience": Paul Cleary, "Miner Tried to Fudge Study: Anthropologist", *The Australian*, 17 October 2012.
"It is not uncommon for us to raise queries": Cleary, "Miner Tried to Fudge Study".
"Lillian Maher did not declare": Paul Cleary, "Native Title Boss Didn't Reveal Ties with FMG", *The Australian*, 17 November 2012.
"Kangeenarina Creek has a story associated": Department of Indigenous Affairs, briefing note to Minister for Indigenous Affairs, 23 December 2011.
"FMG's compliance with the AHA has been variable": Department of Indigenous Affairs, memorandum, James Cook to Cliff Weeks, 1 February 2012.
"Collier denied that his duty to protect indigenous heritage": Neale Prior, "Minister Denies Favours for Mate", *The West Australian*, 4 June 2011.
"unduly influenced by your friendship": "WA Minister Clams Up on FMG Complaint", *AAP*, 21 December 2011.
"a weak and morally wrong decision": "WA Minister Clams Up on FMG Complaint".
"unless she agreed to delete sections": Sue Singleton, letter to Registrar of Aboriginal Sites, 5 November 2011.
"a 'local' archaeological firm could better provide": Singleton, letter to Registrar of Aboriginal Sites.
"The heritage department is very focused": author interview with anonymous former Fortescue consultant, 2013.
"We have spent many millions of dollars to protect": "Fortescue Rejects Claims of Unlawful Heritage Dealing", media release, 7 November 2011.
"Under the deal with Anaconda": Duncan Graham, "Anaconda in Historic Deal with WA Aborigines", *The Age*, 4 April 1997.
"It shows we can share": Graham, "Anaconda in Historic Deal with WA Aborigines".
"That man has a lot to answer for": Marcus Priest, "Brand New Day?" *The Australian Financial Review Magazine*, 28 July 2006.
"They have promised the world": John Flint, "Aborigines to Sue Miner", *The Sunday Times*, 28 April 2002.
"the new deal removed significant heritage": Priest, "Brand New Day?"
"I didn't know what was going on": Andrew Trounson & Kevin Andrusiak,

"FMG's Native Title Deal Unravels", *The Australian*, 13 August 2005.
"They had us in a tight corner": Trounson & Andrusiak, "FMG's Native Title Deal Unravels".
"pushed their way into our offices": Priest, "Brand New day?"
"constantly being pushed and pulled": Yamatji Malpa Aboriginal Corporation, media release, 18 July 2008.
"While mining companies make billions": Yamatji Malpa Aboriginal Corporation, media release, 15 October 2009.
"the process of negotiating": Yamatji Malpa Aboriginal Corporation, media release, 4 June 2010.
"It amazes me to hear": "Iron and Dust", *Four Corners*.
"They want to force upon indigenous people": author interview with Simon Hawkins, 2013.
"If he didn't have a mining company": author interview with Michael Woodley, 2013.
"he has delivered on all his promises": author interview with Clinton Wolf, 2013.
"When I was a kid": author interview with Clinton Wolf, 2013.

CHAPTER 14: DOING GOD'S WORK
"One of the reasons I became a Christian": Andrew Forrest, speech to Pastoralists and Graziers Association of Western Australia convention, September 2012.
"I get all the inspiration": Paul Lampathakis, "Twiggy's Big Leap of Faith", *The Sunday Times*, 2 May 2010.
"He wasn't really religious": author interview with anonymous friend of Andrew Forrest, 2013.
"because that is what gives life to people": Nicola Forrest chats to Debbie Gould, *Deb's Place*, TV program, episode 5.
"What all my kids know": Drummond, "Forrest Fire".
"If this acts as an accelerant": Andrew Burrell, "Miner Shows Charity Mettle", *The Australian Financial Review*, 21 September 2007.
"If the options had been issued": Neil Chenoweth, "Charity Begins at Home for Forrest", *The Australian Financial Review*, 19 August 2010.
"The ACT did indeed receive a huge donation": Neil Chenoweth, "Charity Begins at Home for Forrest".
"As I was driven into work": Forrest v Commissioner of Taxation [2010] FCAFC 6 (5 February 2010).
"When the media revealed the nature of the deal": Peter Klinger, "Glencore's $3.5m Anaconda Squeeze", *The Australian Financial Review*, 11 June 2002.
"These were 'administrative oversights'": John Phaceas, "Forrest Gets Trust Cash", *The West Australian*, 8 June 2002.
"Don't we all do something for our personal tax gains": "The Mine Field", *Four Corners*.

"Forrest said he was hurt by criticism": "The Mine Field", *Four Corners*.
"The donation that Andrew sought": "The Mine Field", *Four Corners*.
"If we were too sensitive": *Australian Philanthropy*, Issue 81, May 2012.
"to call the indigenous employment pact a 'covenant'": Nicola Forrest chats to Debbie Gould, *Deb's Place*, TV program, episode 5.
"50,000 jobs in two years": author interview with anonymous source, 2013.
"It was a complete hit to the solar plexus": Sarah Elks & Patricia Karvelas, "Twiggy Eyes 50,000 Jobs", *The Australian*, 4 August 2008.
"This is the Australian Employment Covenant": Elks & Karvelas, "Twiggy Eyes 50,000 Jobs".
"verging on being slightly dangerous rhetoric": "Indigenous Jobs, Time to Fix Great Wrong: Rudd", *PM*, ABC Radio, 30 October 2008.
"claiming Rudd had not followed through": Andrew Burrell, "Twiggy Ups Challenge in Indigenous Jobs", *The Australian*, 30 April 2012.
"represented a notable shift in the landscape": K. Jordan & D. Mavec, *Corporate Initiatives In Indigenous Employment: The Australian Employment Covenant Two Years On*, ANU, working paper 74 of 2010.
"I've shared with the prime minister": "Forrest Stands by Indigenous Employment Scheme", *PM*, ABC radio, 4 August 2008.
"The problem is structural": author interview with Warren Mundine, 2013.
"I just wanted to drag myself out of bed": author interview with Warren Mundine, 2013.
"A lot of people say all sorts": author interview with Warren Mundine, 2013.
"I was blown away": author interview with David Collard, 2013.
"Cate Blanchett ... whom he described as a 'mate'": Grant McArthur & Cheryl Critchley, "Cate Brings Ray of Hope", *Herald Sun*, 30 March 2009.
"In a recession, in a smashed economy": Nick Lenaghan, "Sounds of Support Touch Stricken Town", *The Australian Financial Review*, 29 June 2009.
"Forrest is one of the people": Paul Barry, "Twiggy's Bushfire Assistance 'inspiring'", *The Power Index*, 14 March 2012.
"He told us to take a ride in the chopper": Barry, "Twiggy's Bushfire Assistance 'inspiring'".
"I had one of Australia's richest men": Julie-Anne Davies, "Magnate Donates Homes to Fire Town", *The Australian*, 28 February 2009.
"I couldn't fault him": author interview with John Brumby, 2012.
"while Forrest's generosity was appreciated": author interview with anonymous Marysville community leader, 2013.
"The Victorian taxpayer was left with a bill": author interview with anonymous Marysville community leader, 2013.
"Twiggy's a very complicated individual": author interview with anonymous Marysville community leader, 2013.
"They [the shares] are dividend-creating": Stephen Bevis, "Miner Gives Millions to Support Art Endeavours", *The West Australian*, 24 August 2011.

"I laughed and said, 'Mate'": author interview with Warren Mundine, 2013.
"fastest-growing global online movement ever": email to author from Nick Grono, 2013.
"We're able to reach into the most influential families": Nick Evans, "Miner's Next Mission is to Tackle Slavery, *The West Australian*, 3 December 2012.
"Andrew and Nicola's 'amazing leadership'": Julia Gillard, speech at International Women's Day breakfast, 8 March 2013.
"his boss is highly dedicated": email to author from Nick Grono, 2013.
"his most inspiring historical figure": Anthony DeCeglie, "Don't Write WA Off", *The Sunday Times*, 7 July 2013.
"Forrest's favourite book": DeCeglie, "Don't Write WA Off".
"'the vast majority' of their fortune": Andrew and Nicola Forrest, The Giving Pledge, 2013.
"With laughter we read": Andrew and Nicola Forrest, The Giving Pledge, 2013.
"Charity is not a substitute for paying tax": James Massola, "Twiggy to Fight World Slavery", *The Australian Financial Review*, 2 June 2012.
"He's not unaware of how philanthropic issues": author interview with anonymous former Fortescue executive "F", 2013.
"I genuinely believe he wants to do good": author interview with Warwick Grigor, 2013.
"He does not want people": Michael Sainsbury & Russell Skelton, "Twiggy's Big Giveaway", *Good Weekend*, 30 March 2013.
"one of the world's greatest philanthropists": author interview with John Poynton, 2013.

### CHAPTER 15: WEALTH AND POWER
"I still walk around barefoot": Jacobs, "$2 Billion Man Seizes the Moment".
"Once you can look after your own needs": Stewart, "The Accidental Billionaire".
"they dined on quail": Knight, "Lunch with Andrew Forrest".
"you have a perfectly good boat": Stewart, "The Accidental Billionaire".
"Shoot me if I ever get a private jet": author interview with anonymous friend of Andrew Forrest, 2012.
"When I first met him": author interview with Rodney Adler, 2013.
"During a tour of the station": Katherine Fleming, "Forrest at Home in Red Dust", *The West Australian*, 30 April 2012.
"To see the look on my brother's face": Fleming, "Forrest at Home in Red Dust".
"the size of a small town": WA Mining Warden's Court decision, FMG Chichester v Rinehart and Hancock Prospecting, 24 August 2007.
"It was difficult to fathom": FMG Chichester v Rinehart and Hancock Prospecting.
"this proposed condition by Forrest": FMG Chichester v Rinehart and Hancock Prospecting.

"We are disappointed that someone": Nick Evans, "Sage Seeks Mineral Claim on Forrest Land", *The West Australian*, 6 October 2012.

"These ports and lines will have lives way beyond ours": Tony Barrass, "Billionaires Cheer as Twiggy Ships His First Ore", *The Australian*, 16 May 2008.

"material inefficiencies": FMG submission to the Economic Regulation Authority, 5 July 2013.

"his railways would be open": Fortescue Metals Group, annual report, 2004.

"Fortescue demanded up to $576 million": "Fortescue to Charge Miners up to $554 mln/yr for Pilbara Railway", *Reuters*, 30 May 2013.

"He has both the influence and resources": author interview with Mark Caruso, 2013.

"you might as well be prime minister": author interview with Albert Wong, 2012.

"He has a need to do what his forefathers": author interview with Warwick Grigor, 2013.

"Twiggy won't be happy with": Matt Chambers, "Twiggy's Philanthropy Set to Make a Huge Difference", *The Australian*, 4 June 2011.

"But I think it would kill him": author interview with anonymous friend of Andrew Forrest, 2013.

"he will fall victim": author interview with Brian Burke, 2013.

# INDEX

Adams, Harry 237
Adler, Larry 56
Adler, Rodney ("Rocket")
   Anaconda Nickel 78–79, 83, 92
   FAI 59
   Far East Capital 71
   Forrest and 63–64, 190
   Forrest and wealth 266–267
   HIH Insurance 92
   International Alpaca Management 68, 69
   Jacksons 56, 59
Administrative Appeals Tribunal 248
AEC *see* Australian Employment Covenant
Air India 65
Aitken, Charlie 153
Allied Mining and Processing *see also* Fortescue Metals Group
   Forrest as chairman 103–104, 169–170
   Forrest's deal with Caruso 108–110
   Forrest's plan 105–106
alpaca breeding 67–69, 70
Alpha Archaeology 232
Altman, Professor Jon 251
Ambrecht, Ken 108
America's Cup (1983) 49
Anaconda Nickel 69 *see also* Murrin Murrin project
   Anglo American 96–98, 247
   board 80, 97
   Centaur Mining 92–94
   Central Bore tenements 82–84
   float 76–77
   Fluor Daniel, sues 90
   Forrest and native title claimants 235–236
   Forrest as chief executive 74–75, 80, 263
   Forrest as chief executive stands down 98–99
   Forrest resigns as deputy chairman 1, 102

   Forrest's financial difficulties 100–102
   Forrest's hostile takeover bid 99
   Forrest's plan 78–79, 80
   Forrest's remuneration 92
   Forrest's termination payment 247–249
   Forrest's "three-nickel province" 80–82, 92
   Glencore control 98, 99, 247
   Glencore joint venture 84–85, 88, 96–97, 98
   Johnston as chief executive 99–100
   Lill 55
   listing 80
   Matheson Motteram Projects 95
   as Minara Resources 99–100
   Motteram 95–96
   plainting 118
   Rich stake 86, 97
   scepticism about 79
   share ownership dispute 74–75
   share price 89, 91, 247
   staff retrenchments 100
   Thompson 72–73
   US capital market raisings 85–86, 88, 94, 126
Anglo American 96–98, 247
"animal spirits", Keynes' theory 167
Argus, Don 116
Armstrong, Paul 91
Art Gallery of Western Australia 257
*The Art of War* (Sun Tzu) 97
Asian immigration to Australia 16–17
ASIC *see* Australian Securities and Investments Commission
Association of Mining and Exploration Companies 198
ASX *see* Australian Securities Exchange
Athletics Australia 91, 107
Atlas Iron 202
*The Australian* 195
Australian Children's Trust
   2010–11 donation campaign 257–258

Forrest establishes 245–247
  purchase of Anaconda shares 247–249
  Queensland floods (2011) 254
Australian Constitution 15
  1967 referendum 31
  MRRT as breach 211–212
Australian Employment Covenant 197, 224, 251–253
*Australian Financial Review* 167, 177–178
Australian Labor Party 122
Australian League of Rights 71–72
Australian National University 251, 252
*Australian Philanthropy* 249
Australian Securities and Investments Commission
  Federal Court full bench hearing against Forrest and FMG 183–184
  Federal Court hearing against Forrest and FMG 179–183
  FMG, investigation into 74, 141, 157, 169–170
  FMG and "binding contracts" 127, 171, 173–174, 178
  Forrest, insider trading investigation into 178–179
  Forrest and Leaping Joey 248
  Forrest offers help to 188
  Heffernan 101
  High Court hearing against Forrest and FMG 185–187
  legal costs 187
  Lill 55
  Mallesons 187
  Rowley, insider trading investigation into 171
  Salter 76
Australian Securities Exchange
  FMG 74, 169–170
  FMG and "binding contracts" 178
  FMG and Consolidated Minerals 170
  FMG and continuous disclosure 172–173
  FMG and Hope Downs 170
  FMG reporting 171–172
  Herb Elliott 179
Australian Selection 77
Australian Submarine Corporation 87
Australian Taxation Office 247, 248

Baosteel 140, 156
Barham, Nick 70
Barnett, Colin 125–127, 194
Bartlett, Peter 267
Bartley, Scott 201
Barton, Edmund 19
Baxt, Bob 187
Baxter, Erica 70
BBY 119, 153
BC Iron 163, 202, 270
Beazley, Justice Margaret 68–69

Bechtel Pacific 30
Bell Resources 63
Bellevue Hill (Sydney) 73
Bennetto, Peter 80
Benney, Bruce 48
Benney Partners 48
Berrima (NSW) 265
BHP Billiton 30, 79
  Chinese steel mills 139–140
  FMG and investment banks 118–119
  Forrest, as catalyst for his success 136
  Forrest proposes to share infrastructure 116–117
  funding for Pilbara expansions 160
  iron ore deposits in Pilbara 104, 112–113, 115–116
  iron ore sales to Japan and Korea 168
  MRRT 205, 206, 208, 212
  railways in Pilbara 105
  record in Pilbara 124–125
  refuses to share infrastructure 119–120, 123, 271
  RSPT 194, 195, 202–203, 204–205
  State Agreement Acts 124
  Woodstock Abydos 130
the Bible (New Testament) 242–243, 261
billionaire guilt 262–263
Binsted, Paul 210
Bitar, Karl 205
Black, Ian 220–221
Black, Scotty 23–25
Black Saturday bushfires (Vic, 2009) 254–257
Blanchett, Cate 255
Bligh, Anna 194
Boao Forum for Asia 273
Bond, Alan
  America's Cup (1983) 49
  Bell Resources 62–63
  Burke and 52–53
  as deal-maker 136
  Intersuisse 63, 67
  Kirke Securities 51
  Robe River 105
Bond Corporation 54, 63
Bower, Ron 227
Bowler, John 128–134
Brand, Sir David 31, 32
Braysich, Jeff 55, 56, 57, 189–190
Brazilian tax rates 195
Brethren 65
British Empire slave trade 260
Brockman Mining 270–271
Bromby, Robin 175
Bronte Capital 160
Broome, Frederick 12
Brown, Clive 124
Brownie, Justice John 94–95
Brumby, John 256
*BRW Rich List* 29–30, 151, 158, 191

Buffett, Warren 245, 261
Bulong project 79
Burke, Brian
　ban on contact with government 122
　CCC 134–136
　FMG lobbyist 124–134
　FMG meeting 123
　Forrest and 273
　jail stints 121
　lobbyist with Julian Grill 121–122
　premier Western Australia 52–53
　State Agreement Acts for FMG 123–125
Burke, Edmund 211
Burrup Peninsula 129–130
business interests and slave labour 259–260
Byford (Perth) 265

*Caijing* 182
Campbell, Ian 135
Campbell, James 97
Campbell, Rod 162, 189
Canadian tax rates 195
Canning, Sadie 235, 248
Cape Lambert iron ore plant 30
capesize vessels 149
Carnarvon Hostel 22
Carpenter, Alan 128, 132, 134
Caruso, Mark 108–110, 272
Castro, Fidel 77, 86
Catlow, Chris 106, 142, 170, 176, 188
Cawse project 79, 92
CCC *see* Corruption and Crime Commission
Ceaușescu, Nicolae 86
Centaur Mining 92–94
Central Bore 82–84
CEO Sleepout 257
Chanos, Jim 158–159, 164
Charlesworth, Ric 34
Charlton, Andrew 203
Cheedy, Ned 227
Chenoweth, Neil 246
Chichester expansion project 196
Chichester Range 112–113, 118
　native title agreements 236–237
child labour 162
China 71, 174, 177, 273
China Harbour Engineering Company 174–175
China Metallurgical Construction Corporation 174–176, 177, 180, 182
China Railway Engineering Corporation 127, 174
China Railway Materials 177
Chinese in Australia 17
Chinese steel mills 111
　bullying by BHP, Rio and Vale 139–140
　crash in iron ore demand 155

iron ore contracts with BHP 117–118
Maanshan conference 170–171
prepayment for FMG iron ore 137–140
resources super cycle 144
resources super cycle slowdown 158–160, 274
Chinese walls 56
Christ Church Grammar (Perth) 34–39
Christian Alternative Movement 71–72
Christianity 242–243, 261
Christians in the Marketplace 243
Christmas Creek 113, 115, 156, 236, 268–269
Churchlands Senior High School 42
CIBC 94
Circular Quay apartment 265
Citigroup 142, 143
Clark, Phil 268
Clarke, Neil 115
Clinton, Bill 86
Cloudbreak 114–115, 118, 150, 153, 156, 236, 268–269
Clout, John 112–113, 119, 137, 142
coal 205
Cocos Islands 68
Coe, Sebastian 107–108
Collard, David 253–254
Collier, Peter 232, 233–234
Commonwealth Day (2012) 1, 262
Connell, Gary 267
Connell, Laurie ("Last Resort")
　borrowing from 49–50
　Burke and 52–53
　Forrest and 61
　Rigoll and 54
Consolidated Minerals 170
Coogan, Allan 96
Cook, James 232–233
Cook, Joseph 19
Coolum conference (Qld) 250–251, 252
Coppin, Brian 63, 67
Coppin, Peter 23
Corporations Act 171, 173, 181, 184, 187
Corruption and Crime Commission
　Bowler 129, 131, 132, 134
　Burke and Grill 122, 134–136
　Forrest–Grill response to Woodstock Abydos decision 132–134
Costello, Peter 135
Cottesloe (Perth) 85, 100, 106, 111–112, 264–265
County NatWest Securities 70
Court, Charles 32, 123
Court, Richard 78, 126
Court of Appeal, Victoria 94
*Crikey* 164
Crowley, Frank 200
Crown Casino 253
Cuba 77
Cumming, Ian 141, 159, 165–166

Cyclone George 148

Dampier 216, 217
Davies, Valerie 248
Deakin, Alfred 19
Dean, Steven 95
Dennis, Stephen 93
Department of Indigenous Affairs
   criticism of indigenous training 253
   FMG 232–234
   Roebourne 217
Diggers and Dealers 108, 151, 212
disaster relief 254–257
Dixon, Kris 222
Dobinson, Ross 58, 62
Drage, Raymond 237
Drayton, Ken 244
Drummond, Mark 245
Ducie, Paul 83
Durie, John 166

Economic Regulation Authority (WA) 271
Eddington, Rod 34, 250
Edgar, Bill 44
Edwards, Damon 115
Edwards, Malcolm 65
EIE International 64
Elizabeth II 1, 262
Elliott, Herb 108, 116, 179, 209–210, 219–220
Elliott, John 180
Emerson, Craig 195
Emichrome 151
Emmett, Arthur 183
Ensor, Ben 68, 69
Equity-1 101
Eureka Heritage History Archaeology 234–235
Evans, Gareth 259
Eyre, Edward John 10

Facebook 259
FAI 59 *see also* Adler, Rodney ("Rocket")
Fairbridge 257
Fairfax Media 273
Falcone, Phil 140–141
Far East Capital 70–71, 73–75, 76
FBI 86
Federal Court 144, 173, 179–183, 229
Federal Court full bench 183–184, 248
female suffrage 15–16
Fengli Group 152
Ferguson, Martin 194–195, 204–205
Finkelstein, Ray 183
Fisher, Wei 176, 182
Flanagan, David 202
Fluor Daniel 87–88, 90, 98
FMG *see* Fortescue Metals Group
Fogarty, Brett 51
Forrest, Alexander 10, 17–18, 25

Forrest, Andrew *see* Forrest, John Andrew Henry ("Twiggy")
Forrest, David (brother) 8, 28, 33, 71
   Christ Church Grammar 36, 37–38
   Forrest gives Fortescue shares to 246
Forrest, David (great-grandfather) 17, 25–27
Forrest, Don (father) 7, 23, 237
   Forrest, management of Minderoo 45–46
   Forrest, relationship with 28
   Forrest and *Geordie Girl* (boat) 265–266
   Hale School 40
   manager of Minderoo 27–28, 45
   Mervyn's treatment of 27
   Nicola Maurice 66
   stuttering 35
Forrest, Don (great-uncle) 27
Forrest, Grace (daughter) 73, 189, 245, 258–259
Forrest, Ivan 235–236
Forrest, Janie *see* Hicks
Forrest, Sir John
   attitudes to Aborigines 16
   attitudes to Asian immigration 16–17
   attitudes to women 15–16
   career 8–9
   death at sea 19
   education 25, 40
   executive council member 12
   federal portfolios 18–19
   federation 15
   Forrest, influence of legacy on 6–8
   Leichhardt expedition (1869) 9, 80, 115
   Liberal Party, bid for leadership 19
   marriage 12
   premier Western Australia 13–14, 272
   public works 13–14
   Rudd invokes 199–200
   surveying expedition (1870) 10
   transcontinental expedition (1874) 10–12
   transcontinental railway (1917) 19
Forrest, John Andrew Henry ("Twiggy")
   as Andrew 6
   birth 5–6
   childhood Aboriginal companions 22–23
   childhood at Minderoo 21–22
   Christianity 242–243, 261
   dress 47–48, 51
   education: Carnarvon Hostel 22
   education: Christ Church Grammar 34–39
   education: Hale School 39–44
   education: Onslow Primary School 22
   education: School of the Air 22
   education: UWA 43, 45
   employment: Cape Lambert 30

employment: Jacksons' Perth office 53–57
employment: Jacksons' Sydney office 60, 61–62
employment: stockbroking 48, 50–51
as entrepreneur 3–4, 136 *see also* Anaconda Nickel; Fortescue Metals Group
father, relationship with 28
high profile 191–192
moral courage 51–52, 220
mother, relationship with 29
nicknames: "Kookaburra" 61
nicknames: "Silver Tongue" 45, 50–51
nicknames: "Twiggy" 36
Nicola Maurice: marries 71, 244
Nicola Maurice: meets 65–66
as pastoralist 268–270
as philanthropist *see* philanthropy
as politician 213–214, 271–274
politicians and 262
private jet 266
stuttering 34–35, 39, 42–43, 47
truthfulness 189–190
wealth 244–245, 264–267
Forrest, Judy (mother) 5–6, 22, 28, 32–33, 58, 68, 152
Forrest, Lily (great-aunt) 27
Forrest, Mary 26
Forrest, Matilda (daughter) 244
Forrest, Mervyn (grandfather) 27
Forrest, Nicola (wife) 108, 111–112
 Australian Children's Trust 257
 family background 71
 Forrest, meets and marries 65–66, 244
 Giving Pledge 260–261
 Leaping Joey 248
Forrest, Sophia (daughter) 84, 188, 189, 245
Forrest, Sydney (son) 40, 100, 189, 245
Forrest, Violet (great-aunt) 27
Forrest & Forrest 269
Fortescue Metals Group 268–269
 asset sale 163
 Australian investment banks 118–119
 Australian investors 153–154
 BC Iron equity 270
 board 107–108, 156, 159
 bond issue 142–143
 Bowler 128–134
 Burke and Grill: lobbyists 123, 124–136
 Chinese equity demands 175, 177–178, 180, 182, 188
 customers *see* Chinese steel mills
 exploration campaign 112–116
 extent of holdings 116
 feasibility study 126
 Forrest and possible disqualification as director 173–174

Forrest as chief executive 103–104, 169–170
Forrest as chief executive stands down 157
Forrest as non-executive chairman 157
Forrest proposes to share infrastructure 116–117
Forrest's deal with Caruso 108–110
Forrest's management style 145–149, 157
Forrest's plan 105–106
Forrest's response to resource rent tax *see* resource rent tax
Forrest's stake 152–153, 157–158
Forrest's stake and philanthropy 261
Forrest's stake, increase in 159–160
Hunan Valin equity 156
indigenous business contracts 222–223
indigenous relations group 161
indigenous royalties *see* native title agreements
indigenous workers 221–222, 254
infrastructure 144, 201
infrastructure budget 145, 149
infrastructure contracts 141, 174–176
infrastructure: surface miners 118, 145
investors 140–142
loan covenants 164, 165–166
loan funds 164–165
maiden shipment 150
name 104
Njamal joint venture 223
office clock 148, 150
office in Perth 112, 145–146
Pilbara dongas 256–257
plaints over BHP ground 118
production targets 158, 159, 163, 168
railway completion 148, 149–150
railway contract 127, 174
railway in Pilbara 111, 117
railway infrastructure 150
railway open-access regime 270–271
railway through Woodstock Abydos 129, 130–134
regulation *see* Australian Securities and Investments Commission; Australian Securities Exchange
revenues 156–157
share donations 257–258
share price 109, 151, 155, 156, 164
share price after RSPT announcement 196
share price shorted 158–159
shareholder dividend 157
staff 145–149
staff retrenchments 161–163, 168
staff share options 147
State Agreement Acts 123–125, 135, 270
State Agreement Bill 125–127

taxes 198–199, 208, 246–247, 261–262
Fortescue River 104
Fox, Lindsay 251, 256
Frazer, Andrew 266
Freeman, Cathy 91
Freeport 77

Gallagher, Michael 231–232
Gallop, Geoff 122, 125, 128
Gates, Bill 245, 258, 261
GenerationOne 249, 253, 258
*Geordie Girl* (boat) 265–266
Gibson, Belinda 188
Gillard Labor government
    2010 federal election 207–208
    Forrest's challenge to MRRT in High Court 211–212
    Forrest's slavery campaign 260
    leadership coup against Rudd 203, 204, 206, 212
    negotiates MRRT 205
    truce with big miners 204–205
Gilmour, Justice John 180, 181–182, 183
Glencore 75, 84–85, 86, 88, 96–97, 98, 99, 247
global financial crisis (2007–08) 155–156, 193, 246, 247, 255, 266
Goldberg, Yosse 56
Goldfields *see also* Murrin Murrin project; Pilbara
    gas pipeline to 78
    Jacksons' trips to 58–59
    mineral wealth of 9–10
    water pipeline to 7, 8, 13, 81
Goode, Brad 230–231, 232, 234
Goodyear IV, Charles ("Chip") Waterhouse 117–118
Gore, Mike 65
Goyder, Richard 40–41
Gray, Gary 195
GRD 51
Great Australian Bight 10
Great Barrier Reef 64
Greenvale project 79, 84
Grigor, Warwick 6, 70, 71, 72, 73–75, 119, 249, 262–263, 272
Grill, Julian
    ban on contact with government 122
    CCC 134–136
    FMG lobbyist 124–134
    FMG meeting 123
    lobbyist with Brian Burke 121, 122
Grill, Lesley 135
Grono, Nick 259, 260
Gutnick, Joseph 56, 92–94

Hagan, Stephen 230
Hale School (Perth)
    Forrest at 25, 39–44
    Forrest Library 40

Hale School Foundation 41
Halls Creek 219–220
Hamersley, Edward 115
Hamersley, Margaret 12, 16, 115
Hamersley Iron 30
Hamersley Ranges 29, 112–113, 115–116, 215
Hancock, John 106–107
Hancock, Lang 29, 32, 106, 225
Hancock Prospecting 106–107
Handler, Richard 143
Hannon, David 63
Hannon, Eamon 112, 113–115, 137, 162
Harbinger Capital 140–141
Harken, Ron 267
Hartley Poynton 47
Hawke Labor government 48, 49, 195
Hawkins, Simon 239
He Lianzhong 180–181, 182
Heathrow Airport explosives 143
Heenan, Justice Eric 95
Heffernan, Shayne 101–102
Hempton, John 160
*Heng Shan* (ore carrier) 150
Henry, Ken 193
Henry VIII 260
Hewson, John 73, 80
Heydon, Dyson 186
Heyting, Ed 147, 181
Hicks, Aldous 80
Hicks, Janie (sister) 7, 28, 63
    Forrest gives Fortescue shares to 246
    International Alpaca Management 68, 69
High Court of Australia
    Anaconda–Centaur deal 94
    FMG *versus* Puutu Kunti Kurrama and Pinikura native title group 238
    FMG's "binding contract" 185–187
    Forrest's challenge to MRRT 210–212
    Mabo 221, 226
    native title as property right 238–239
high-pressure acid leaching (HPAL) 77–78, 88
HIH Insurance 92
Hill, Julian 63, 66–67
Holmes à Court, Peter 34
Holmes à Court, Robert 56, 136
Hope Downs 170
Howard, John 89–90, 97, 111, 135, 175, 176, 243
Howarth, Ian 177–178
Human Rights and Equal Opportunity Commission 71, 72
Hunan Valin 156
Huston, Peter 176, 182, 183, 185, 188

indigenous Australians
    1967 referendum 31
    FMG business contracts 222–223

INDEX  305

FMG indigenous relations group 161
FMG workers 221–222, 254
Forrest and creation of indigenous jobs 197, 244, 249–254
  labour force statistics 251
  mining industry overlooks 31
  mining royalties *see* native title agreements
  Noonkanbah dispute 32
  pastoralists sack 31–32, 219
  Pilbara rock engravings 129–130
  Sir John Forrest's attitudes to 16
  stockmen's equal pay ruling (1965) 31, 219
  stockmen's strike (1946) 23–25
International Alpaca Management 68–69, 70
International Crisis Group 259
Intersuisse 63, 64–67
Iran 86
iron ore
  2003 price 103, 110
  2006 price 153
  2008 price 151
  2010 export value 192
  2012 price 168
  2013 price 3
  deposits in Pilbara 29, 103–104
  forcing up of price 139–140
  Forrest's global demand prediction 103–104
  Joint Ore Reserves Committee (JORC) code 172
  microplaty haematite 112–113
  mining 144–145
  MRRT 205
  surface mining 118, 145
Iron Ore Holdings 225
Irving, David 72
Israeli invasion of Lebanon 143

Jackson, David, QC 211
Jacksons
  collapse of 62
  Forrest manages Perth office 53–57
  Forrest transfers to Sydney office 60, 61–62
  Forrest's shares in 62
  trips to WA Goldfields 58–59
James, Mal 254
Japanese steel mills 138, 167–168
Jefferies 143
Jeffries, Nyaparu 238
Job Services Australia 252
Johnson Winter & Slattery 187
Johnston, Peter 99–100, 236
Joint Ore Reserves Committee (JORC) code 172
JP Morgan 119, 196
Judge, Bruce 65

Kaiser Steel Corporation 30
Karkar, John, QC 180, 181
Karratha 216–217, 223
Kazakhstani oil venture 54
Keane, Chief Justice Patrick 183
Keating, Paul 121, 165
Keep Australia Afloat 207
Keeves, John 187
Kennedy, Ray and Murray 113
Kepert, Doug 115
Keynes, John Maynard 167
Kimberley region 18, 32, 219–220
Kings deposit 168
Kirchlechner, Philip 111, 138, 147, 167–168, 170
Kirke, Graeme 7, 50–51
Kirke Securities 7, 50–51, 53
Kloppers, Marius 195, 208
Knight, Barry 162, 170
Knight, Elizabeth 265
Korean steel mills 168
Kovacs, Akos 36

Langton, Marcia 223–224, 250–251
Lawcock, Glyn 158
Lawrence, Magistrate Bob 83–84
Lazard 165
Leaping Joey 247–249, 263
Lebanon, Israeli invasion of 143
Lee, Jeremy 72
the Left 224
Lehman Brothers 155
Leichhardt, Ludwig 9
Leucadia International Corporation 141, 159, 165–166
Li Xiaowei 156
Liberal Party 19, 272
Liberal-National Coalition 207
Lill, Simon 54–55, 59
Lilleyman, Greg 119
Linegar, Chris 108
Liu, David 137–140
Liu Guoping 177
London, Tony 35, 37–38
London *Times* 151
Lowry, Ann Marie 162, 189
Lowy, Frank 251
Lucy, Jeff 173

Ma Yanli 176
Mabo 221, 226, 238–239
McComb, David 34
McCusker, Malcolm, QC 50
Macdonald, George 83
McGlew, Blair 228
McHale, Sheila 130–134, 135
McKew, Maxine 194–195
Macquarie Bank 196
Macquarie Equities 153

MacTiernan, Alannah 132
Maher, Lillian 231–232
Maher, Lisa 231
Maish, Richard 94
Mallesons 187
*Man For All Seasons, A* (Bolt) 260
Mandela, Nelson 66
Margaret River 114, 266
Markland House 63, 64
Marlborough, Norm 125, 126–127
Martin, Wayne 34
Marysville (Vic) 254–257
Masterman, Isaac Murrin 87
Masterman, Michael 87
Matheson, Peter 84, 95
Matheson Motteram Projects 95–96
Matlin Patterson 99
Maurice, Katrina 72
Maurice, Nicola *see* Forrest
Maurice, Tony and Brooke 71–72
Meares, Jodhi 70
Melbourne Football Club 92
Mendelawitz, David 112, 113, 114, 137, 149, 153–154, 190
Meredith, Maria 236
Meurs, Peter 145, 234–235
Meyer, Donna 238
MGA Consulting 231–232
Millstream-Chichester National Park 215
Minara Resources 99–100
Minchin, Tim 34
Minderoo sheep station 17, 21–28
   under David Forrest's management 45–46
   under Don Forrest's management 27–28, 45
   establishment of 25–27
   Forrest and Christianity 242–243
   Forrest and gas company compensation 225–226
   Forrest and journalists' visit 267–268
   Forrest buys back 267
   Forrest visits with Wong 65
   Forrest's early childhood at 21–22
   leaving for boarding school 32–33
   under Mervyn Forrest's management 27
   mineral wealth 226, 268–270
   Nyiyaparli people 237
   Scotty Black as head stockman 23–24
   stockmen's equal pay ruling (1965) 219
Minerals Council of Australia 192, 198
minerals resource rent tax 205–213
Mining Act of Western Australia 118, 269
mining and two-speed economy 193
mining boom 3, 144
mining boom slowdown 158–160, 274
mining iron ore 144–145 *see also* iron ore
mining nickel laterites 77–78, 79, 88
mining on pastoral leases 268–270
mining royalties
   to native title holders 218, 235–236
   state royalties and MRRT 212
   to WA government 124
mining tax *see* resource rent tax
mining welfare 218
Mission Australia 257
Mitchell, Bill 207
Mittal, Lakshmi 141
Moa Bay (Cuba) 77–78
*The Monthly* 209
Moody's 159
More, Sir Thomas 260
Morgan, Hugh 79
Morgan, Sally 131
Morrison, John 44, 45, 71
Mossad 86
Motteram, Geoff 76, 77, 80, 95–96
Mount Margaret 80
Mount Nicholas 105, 112–113
Mount Tom Price 237–238
Mount Whaleback 30, 31
Moyes, Peter 35–36
MRRT *see* minerals resource rent tax
Mulga Downs pastoral station 107, 268–269
mulgara 269
Mundaring Weir (outside Perth) 7, 13
Mundine, Warren 250–251, 253, 258
Murdoch University's Institute for Immunology and Infectious Diseases 257
Murrin Murrin project *see also* Anaconda Nickel
   autoclaves 86–87, 88–89
   chapel 90, 235
   construction work 87–88
   expansion plan 90–91
   fixed-price contract 87–88
   location and value 77, 78, 80
   native title agreements 235–236
   official opening (1999) 89–90, 96, 97, 243
   plant filters 87
   success 98
   unproductive 97–98
Myers, Allan, QC 180, 181, 188

NAPLAN 217
National Australia Bank 62
National Competition Council 117, 123
National Development and Reform Commission (China) 180–181, 182
*National Indigenous Times* 230
National Native Title Tribunal 229, 231–232, 237–238
Native Title Act 221, 230, 237–238
native title agreements *see also* Woodstock Abydos
   between Anaconda and Murrin Murrin native title holders 235–236

in Chichester Range 236–237
effect of mining money on indigenous groups 240–241
between FMG and Nyiyaparli elders 223, 236–237
between FMG and Puutu Kunti Kurrama and Pinikura native title group 237–238
between FMG and Yamatji Marlpa Aboriginal Corporation 239
between FMG and Yindjibarndi people *see* Yindjibarndi people
indigenous labour *versus* compensation 239
with miners in general 224–225
mining royalties 218, 235–236
native title as property right 238–239
traditional owner register 228
Natural Resource Management 254
Nepalese orphanage 258
*The New York Times* 151
New York Yacht Club 49
Nicholson, Robert 215
nickel laterite extraction 77–78, 79, 88
Nigeria 86
night parrot 107
Nintendo 260
Njamal people 223
Noonkanbah dispute 32
Norman, Greg 255
North West Shelf gas project 129–130
Nyiyaparli people 223, 236–237

Oakeshott, Rob 207–208
O'Connor, CY 7, 13, 14–15
O'Faircheallaigh, Ciaran 239
Officer Basin 81
O'Keeffe, Michael 75, 97
Oliver, Craig 203–204
Olympic Games (Beijing 2008) 197
Olympic Games (London 2012) 107–108
Olympic Games (Sydney 2000) 91, 92
Onslow Primary School 22
O'Reilly, Louise 106, 152

Packer, James 70–71, 150, 151, 251, 253
Packer, Kerry 70, 151
Palmer, Clive 2, 209, 213, 273
pastoralists
    mining on pastoral leases 268–270
    response to indigenous stockmen's equal pay ruling 31–32, 219
    response to indigenous stockmen's strike 23
Pat, John 217
Patersons 55
Pearce, Georgina 58
Pearce, Merrilee 58
Pearce, Stephen 161, 163, 189, 195–196, 200, 201, 203

Pearce, Woody 57–58
Pearson, Noel 250–251
Peepingee Trust 246
Peppermint Grove (Perth) 266
Perth stock exchange 57
Pfafflin, Bob 53, 60
philanthropy 242
    Forrest and Aboriginal disadvantage 249–254
    Forrest and disaster relief 254–257
    Forrest and ending slavery 23, 258–260
    Forrest donates Fortescue shares and Poseidon options 245–247
    Forrest's motivation 261–263
    Giving Pledge 260–261
    role for media 249
Philanthropy Australia 162
Pilbara *see also* Goldfields
    arrival of large-scale mining 29–30, 216
    Cyclone George 148
    extent of FMG's holdings 116, 268–269
    iron ore deposits 29, 103–104
    mine worker salaries 216–217
    mining industry overlooks indigenous labour 31
    Roebourne 216–217
    Yindjibarndi country 215
Pilbara Native Title Service 236, 237
Pinochet, Augusto 86
Pixley, Rohan 36, 41
plainting 118
Port Hedland 114, 149, 150
Porter, Christian 40–41
Poseidon Nickel 245–247
Potts, Malcolm 244
Povey, Mark 56
Power, Nev 157, 161, 163, 168, 188
Poynton, John 47, 56, 263
Priest, Marcus 236
Puutu Kunti Kurrama and Pinikura (PKKP) native title group 237–238

Queensland
    floods (2011) 254
    RSPT 194
    two-speed economy 193

Rankine-Wilson, Adam 44
Ravensthorpe plant 79
regulation *see* Australian Securities and Investments Commission; Australian Securities Exchange
resource rent tax
    on coal 205
    "Dutch disease" 193
    effect on FMG 195–196
    Forrest-Rudd pact 200–204
    Forrest's challenge to MRRT in High Court 210–212

Forrest's response to MRRT 206, 207–210
Forrest's response to RSPT 193–194, 195–196, 197–198
government revenues 192, 205, 206, 210, 212
on iron ore 205
minerals resource rent tax (MRRT) 205–213
Perth anti-tax rally 199
on petroleum products 195
reasons for 192
resource super-profits tax (RSPT) 192–196
Tapp 162
Rich, Marc 86, 97
Richard, Peter ("Cabbie") 54, 61–62
Richards, Wayne 89–90
Richfile Pty Ltd 83
Rigg, Eddie 143–144
Rigoll, Dave 54, 55, 59
Rinehart, Gina 2, 29–30, 106–107, 170, 199, 207, 209, 213, 225, 268–269, 273
Rio Tinto 29, 30
    Chinese steel mills 139–140
    employment of indigenous workers 221
    FMG and investment banks 118–119
    Forrest, as catalyst for his success 136
    Forrest proposes to share infrastructure 116–117
    funding for Pilbara expansions 160
    iron ore deposits in Pilbara 104, 112–113, 115–116
    iron ore production targets 158
    iron ore sales to Japan and Korea 168
    MRRT 205, 206, 208, 212
    railways in Pilbara 105
    record in Pilbara 124–125
    refuses to share infrastructure 119–120, 123, 271
    royalties to Gina Rinehart 225
    royalties to Pilbara indigenous groups 218
    RSPT 194, 195, 202–203
    RSPT: truce with Swan 204–205
    State Agreement Acts 124
Robe River 105
Robertson, John 186–187
Robinson, Michael 232
Robinson, Sir William 13
Roebourne 216–217, 223, 226, 229
Rothwells 50, 54
Rowley, Graeme 105–106, 109, 142, 152, 170, 171, 176, 188
Rowley, Lisa 105
Roy Hill pastoral station 113
Rudd, Kevin 89, 135, 196–197
Rudd Labor government 156
    2010 federal budget 194
    Chinese delegation to 203
    Forrest–Rudd indigenous jobs plan 250–253
    Forrest–Rudd resource tax pact 200–204
    Gillard leadership coup 203, 204, 206, 212
    resource super-profits tax 192, 197, 198, 199–200 *see also* Swan, Wayne

Sage, Tony 269–270
St Vincent de Paul 257
Salter, Peter ("Salty Pete") 76–77, 82–84
Salvation Army 257
Sanders, Jon 34
Savas, Kerry 228–229
School of the Air 22
Schumpeter, Joseph 4
Schwab, Adam 164
Scovell, James 188
Scrimshaw, Russell 108, 138–139, 142–143, 150, 188
Securities and Exchange Commission (SEC) 140–141
September 11 (2001) 1, 101–102
Shanahan, Dennis 195
Shen Heting 177
Sherritt International 78, 88, 97
Simon and Garfunkel 255
Singleton, John 207
Singleton, Sue 234–235
slavery 23, 258–260
Smith, Graeme 267
social media 259
Solomon mining hub 115, 196, 215
    Aboriginal Heritage Act (WA) 230–235
    Alpha Archaeology 232–233
    Kangeenarina Creek 231, 232, 233
    Yindjibarndi people and FMG 218, 224–225, 226–230
Southern Cross Equities 153
Spohler, Bruce 94
Standard & Poor's 142–143
Staniforth, Walter 248
State Administrative Tribunal 233
state mining royalties 212
Steinberg, Joseph 141, 159
Stevens, Matthew 174
Stock, David 237
stockbroking industry
    1983–87 bear market 47–48, 55–56
    1987 sharemarket crash 59–60, 62–63
    regulation 56
Stokes, Kerry 150, 225, 251
Strapp, Bruce ("Jock") 39
Suffrage League 16
Sugar, Eddie 142
Sundance Resources 204
Supreme Court of New South Wales 94
Supreme Court of Victoria 93

Supreme Court of Western Australia 95, 165
Swan, Wayne
　Forrest and 195–196, 208–210
　Forrest and tax 261–262
　Gillard Labor government 204, 206
　MRRT revenues 206, 212
　RSPT: introduction 192
　RSPT: lack of consultation with mining industry 194
　RSPT: lack of consultation with ministers 194–195
　RSPT: pre-emptive allocation of funds 192
　RSPT: truce with big miners 204–205, 210
　Rudd and 194
　Wilkie and 213
Sydney Mining Club 172

*Tales from the Political Trenches* (McKew) 194–195
Tapp, Julian 125, 127–128, 131, 134, 162, 189
Target 260
Thalanyji people 22–23, 25
Thomas, Mark 120, 188
Thomas, Peter 142, 145
Thompson, David 72–73
Tinkler, Nathan 2
Tom Price 237–238
transportation 260
Treasury 201
Tregonning, Ken 40, 41, 43
Tucker, Brian 223
Turnbull, Malcolm 210
2020 Summit (Canberra) 197

Uganda 44–45
University of Western Australia 43, 45
uranium exploration 269–270
US capital market raisings
　Anaconda Nickel 85–86, 88, 94, 126
　FMG bond raising 143–144
　potential default events 196

Vale 139–140
van der Sluys, Stephen 94
Veldhuizen, John 119, 153
Victorian government 256–257
Vimeo 229
Vizard, Steve 180

Walk Free 259
*The Wall Street Journal* 151
*Wall Street* (film) 111
Walsh, Sam 116
Walter, Doug 255
Warden's Court 269
Warren, Justice Marilyn 93–94

Watling, Alan 106, 145, 148
Weatherhead, John 38–39
Weeks, Cliff 232
welfare lobby 219
*West Australian* 91
Western Australia
　2001 state election 122
　Aboriginal rock engravings 129–130
　exploration of 10–12
　federation 15
　mining *see* Goldfields; Pilbara
　mining royalties 124
　RSPT 194
　State Agreement Acts 123–127, 135, 270
　as State of Excitement 48
　transcontinental railway 19
　two-speed economy 193
　WA Inc. 52–53, 56, 121–122
Western Continental 56
Western Mining Corporation 79
Wilberforce, William 260
Wilkie, Andrew 207–208, 212–213
Williams, Daryl 259
Williams, Jim 118
Wilson, Magistrate Stephen 269
Windich, Tommy 10
Windsor, Tony 207–208
Wirlu-murra Yindjibarndi Aboriginal Corporation 226–230, 231, 232
Wolf, Clinton 240–241
Wong, Albert 64–67, 71, 97, 272
Wong, Kie Chie 151
Woodley, Lorraine 216
Woodley, Michael 215–216, 217–218, 224–225, 226–230, 234, 240
Woodside Petroleum 129–130
Woodstock Abydos
　Aboriginal custodian campaigns against 131
　BHP Billiton 130
　FMG diversion possibility ignored 131
　FMG railway through 129, 130–134
　Forrest–Grill response to McHale's decision 132–134
　protection under Aboriginal Heritage Act 129
World Monuments Fund 129–130
World Trade Center 1, 101–102
Worley Parsons 145
Wran, Neville 64
Wren, Chris 69
Wright, Peter 29
Wu Yueming 152
Wyatt, Stephen 177–178
Wylie, John 165–166

Xi Jinping 203, 204
Xstrata 194, 202–203, 204–205, 206, 208, 212

Yamatji Marlpa Aboriginal Corporation 239
Yarri Mining 226, 269
Yindjibarndi Aboriginal Corporation 216, 226–230
Yindjibarndi people
   Kangeenarina Creek 231, 232, 233
   native title agreement between FMG and 218, 224–225, 226–230
   native title rights 215–216
   in Roebourne 217
Young, Mike 202
Young, Neil, QC 180, 181, 183
YouthCare 257
YouTube 229–230
Yungngora people 32

Zaleznik, Abraham 4
Zara 260
Zimpel, Frank 265

www.ingramcontent.com/pod-product-compliance
Lightning Source LLC
Chambersburg PA
CBHW070341240426
43665CB00046B/2322